103 Springer Series in Solid-State Sciences
Edited by Peter Fulde

F. Yonezawa (Ed.)

Molecular Dynamics Simulations

Proceedings of the 13th Taniguchi Symposium
Kashikojima, Japan, November 6–9, 1990

With 102 Figures

Springer-Verlag
Berlin Heidelberg New York
London Paris Tokyo
Hong Kong Barcelona
Budapest

Professor Dr. Fumiko Yonezawa
Keio University, Hiyoshi 3-14-1,
Kohoku-ku, Yokohama 223, Japan

ISBN-13:978-3-642-84715-8 e-ISBN-13:978-3-642-84713-4
DOI: 10.1007/978-3-642-84713-4

Library of Congress Cataloging-in-Publication Data
Taniguchi International Symposium on the Theory of Condensed Matter (13th : 1990 : Kashikojima, Japan) Molecular dynamics simulation : proceedings of the 13th Taniguchi Symposium, Kashikojima, Japan, 6–9, November 1990 / F. Yonezawa, ed. Includes bibliographical references and index.
ISBN-13:978-3-642-84715-8
1. Condensed matter–Congresses. 2. Molecular dynamics–Simulation methods–Congresses. I. Yonezawa, F. (Fumiko) II. Title. QC173.4.C65T36 1990 530.4'1–dc20 91-47017

Softcover reprint of the hardcover 1st edition 1992

Typesetting: Camera-ready by author
54/3140 – 5 4 3 2 1 0 – Printed on acid-free paper

Springer Series in Solid-State Sciences

Editors: M. Cardona P. Fulde K. von Klitzing H.-J. Queisser

Managing Editor: H. K. V. Lotsch Volumes 1–89 are listed at the end of the book

90 **Earlier and Recent Aspects of Superconductivity** Editors: J. G. Bednorz and K. A. Müller

91 **Electronic Properties of Conjugated Polymers III** Basic Models and Applications
Editors: H. Kuzmany, M. Mehring, and S. Roth

92 **Physics and Engineering Applications of Magnetism** Editors: Y. Ishikawa and N. Miura

93 **Quasicrystals** Editors: T. Fujiwara and T. Ogawa

94 **Electronic Conduction in Oxides** By N. Tsuda, K. Nasu, A. Yanase, and K. Siratori

95 **Electronic Materials** A New Era in Materials Science
Editors: J. R. Chelikowsky and A. Franciosi

96 **Electron Liquids** By A. Isihara

97 **Localization and Confinement of Electrons in Semiconductors**
Editors: F. Kuchar, H. Heinrich, and G. Bauer

98 **Magnetism and the Electronic Structure of Crystals**
By V. A. Gubanov, A. I. Liechtenstein, and A.V. Postnikov

99 **Electronic Properties of High-T_c Superconductors and Related Compounds**
Editors: H. Kuzmany, M. Mehring, and J. Fink

100 **Electron Correlations in Molecules and Solids** By P. Fulde

101 **High Magnetic Fields in Semiconductor Physics III**
Quantum Hall Effect, Transport and Optics By G. Landwehr

102 **The Physics of Conjugated Conducting Polymers** Editor: H. Kiess

103 **Molecular Dynamics Simulations** Editor: F. Yonezawa

104 **Products of Random Matrices in Statistical Physics**
By A. Crisanti, G. Paladin, and A. Vulpiani

105 **Self-Trapped Excitons** By K. S. Song and R. T. Williams

106 **Physics of High-Temperature Superconductors** Editors: S. Maekawa and M. Sato

107 **Electronic Properties of Polymers** Orientation and Dimensionality of Conjugated Systems
Editors: H. Kuzmany, M. Mehring, and S. Roth

108 **Site Symmetry in Crystals** Theory and Applications
By R. A. Evarestov and V.V. Smirnov

Participants in the Taniguchi International Symposium

November 6–9, 1990 (at Kashikojima, Ise-Shima National Park, Japan)

Seated (left to right)
D. Frenkel, M. Parrinello, J. Kanamori, F. Yonezawa, S. Yip, D.C. Rapaport, D.J. Evans, J.-L. Barrat, H.C. Andersen

Second row (left to right)
H. Akai, C. Satoko, I. Ohmine, Y. Kaneko, Y. Hiwatari, S. Tsuneyuki, K. Terakura, Ms. Andersen

Third row (left to right)
S. Sakamoto, T. Oguchi, T. Yamaguchi, K. Kawamura, S. Nosé, J. Habasaki, T. Kawakatsu

Preface

This volume contains papers presented at the Thirteenth Taniguchi Symposium on the Theory of Condensed Matter, which was held at Kashikojima (in Ise-Shima National Park), Japan, from 6th to 9th November, 1990. The topic of the symposium was *Molecular Dynamics Simulations*.

The general objective of this series of the Taniguchi Symposia is to encourage developing fields of great promise in condensed matter physics. Our theme, molecular dynamics (MD) simulations, certainly fulfills this requirement, because the field is developing at a remarkable pace and its future is considered almost boundless.

It was in the 1950s that the original idea of the MD methods was first proposed and applied to the study of physical systems composed of many particles. In fact, the invention of the MD techniques occurred soon after the construction of the first computers. For almost 35 years since then, MD methods, together with Monte Carlo methods, have played major parts in the drama of computer simulations.

The triumph of MD simulations is not confined to numerical aspects of detailed analyses of physical systems. MD simulations have verified some unexpected facts and introduced some new concepts, all of which had never been predicted previously from analytical theories. The occurrence of the Alder transition in a system of repulsive particles and the behavior of the long-time tails of the velocity autocorrelation function for a liquid are just two examples of the results achieved by means of MD studies.

Recently, the possibilities of MD simulations have been greatly extended, partly because of some new methods and partly because of the rapid development of computers. As examples of the former, we can name the MD simulations for various ensembles, for nonequilibrium systems and for quantum mechanical particles. When these new methods are applied, MD simulations approximate physical systems still more closely. As for the latter, the progress in computer technology over the last few decades has been simply outstanding. State-of-the-art computers today are characterized by a speed more than a million times faster than that of the early computers in the 1950s. Naturally, the higher the performance of computers, the more elaborate the simulations can be made and the more reliable the results become.

With this situation in mind, it was the purpose of our symposium to study the present status of MD simulations and to investigate possible means to increase the efficiency of the simulations and broaden the applicabilities of the MD techniques. Now that a further improvement in the speed and capacity of computers is well under way, it is quite timely to discuss how we can prepare ourselves to make the best use of future computers. It will afford us great pleasure if these proceedings are able to provide some clues to solving this problem.

In conclusion, we would like to express our sincere thanks to Mr. T. Taniguchi and the members of the Steering Committee of the Taniguchi Foundation for giving us the opportunity to hold the symposium and to publish this volume.

Yokohama, September 1991 *Fumiko Yonezawa*

Foreword

Speech by Toyosaburo Taniguchi

Dr. Kubo, Distinguished Guests and Friends:

The Taniguchi Foundation wishes to welcome the participants to the Thirteenth International Symposium on the Theory of Condensed Matter, who have come from within this country and from different parts of the world. I feel privileged, as a Japanese industrialist, to be given the opportunity of meeting with the world-famous scholars attending this symposium.

The concept of the symposium is unique in that participants, both Japanese and from abroad, are limited to a number suitable for small discussion groups, because of my desire to see them living together, although only for a short period of time, as a close-knit community. It is easy to talk about but difficult to realize the ideal of international friendship and understanding in a world which is steadily growing smaller.

We are deeply indebted to Dr. Kubo and his associates for their efforts, which made this particular symposium possible. Any words of praise should be directed to all those who have lent their support to this program.

So far, I have met in person a total of 287 participants in this Division: 94 from 20 foreign countries and 193 from Japan. Whenever I meet with symposium participants, both young and old, I am always impressed by the unselfishness and sincerity of the dedication they show in the pursuit of their aims. I hope and trust that, even after they have reached the heights of academic fame during the coming decades, they will continue to join forces and help to forge closer bonds of friendship and cooperation in a manner that will make major contributions not only to academia but also to world peace and the welfare of mankind.

In conclusion, we hope that all the participants will return home with warm memories of this symposium and the pleasant times that we have shared.

Thank you. *xxx*

Contents

Part I

Introduction

Introduction to This Volume

F. Yonezawa

Department of Physics, Faculty of Science and Technology,
Keio University, 3-14-1 Hiyoshi, Kohokuku, Yokohama 223, Japan

Abstract. A brief introduction to the molecular dynamics (MD) simulations is given with a view to evaluating the purpose of this volume. A remark is made about the significance of the computer simulations as a method of research in physics. The merits and extents of computer simulations as well as those of the MD studies are discussed. A new trend in the trade of the MD simulations is also mentioned in connection with the articles presented in these proceedings.

1. Computer Simulations

In physics, just like in any other field of natural science, researches have traditionally been carried out both experimentally and theoretically. In fact, natural science, by definition, deals with all phenomena which take place in nature. Observations and experiments reveal the fundamental secrets of nature which underlie the ostensible appearances of these phenomena. On the other hand, theories are evolved in order to clarify the laws which govern nature.

In most cases, theoretical approaches for explaining physical properties are characterized by the following processes:

1. Extract physical rules and factors responsible for the phenomena under consideration.
2. Construct theoretical models by taking these rules and factors into account.
3. Develop, in these models, analytical formulation for the physical quantities to be studied, and solve the formulation rigorously, if possible, or otherwise by means of some approximations.
4. Calculate, if necessary, the numerical values of required quantities on the basis of the final formulae thus obtained.
5. Derive answers or conclusions from the analytical and/or numerical results attained from steps 3 and 4.

Although each of these processes is important in its own right, the first step of grasping the physical grounds is crucial. If one fails in this first step, it is like starting a race in a wrong direction and the goal will never been achieved.

The second and third steps determine the quality of a model. When a very simplified model is adopted, the formulation for that model might allow rigorous solutions, but there are chances that some essential aspects are left out form the model. When one tries to accommodate in a model as many rules and factors as possible, the model would lead to formulation which is not analytically tractable and inevitably has to resort to some approximations.

As for the fourth step, the evaluation of a physical quantity may not be an easy task even when the expression for that quantity is written in a closed form. For example, a numerical value of a quantity may be obtained as a self-consistent solution of some complicated integral equations. In practical calculations, the self-consistency is achieved after a number of iterative steps. This is one of the scenes where computers give full play to their own ability.

There is still another scene, even more exciting and dramatic, where computers play key roles. In this scene, the aforementioned third and fourth steps are replaced by the steps:

3*. Realize a system composed of many particles according to a model constructed in the second step.

4*. Calculate various physical properties of the system thus realized.

In this new approach, the realization of many–particle systems as described in step 3* is named "*computer simulations.*" As practical techniques for performing computer simulations, two major methods have been proposed so far — the Monte Carlo (MC) methods and the molecular dynamics (MD) methods. The time developments of many–particle systems are pursued by means of stochastic procedures in the the MC methods while they are followed by deterministic procedures in the MD methods.

Since the third step of analytical formulation and the fourth step of numerical calculations form the backbone of a theoretical study, the replacement of these two steps by processes 3* and 4* respectively changes the essential features of approaches. In other words, an approach in terms of computer simulations is completely different from any ordinary theoretical approach, though steps 1, 2 and 5 are common to both approaches. In fact, the basic idea of computer simulations is to construct a system which is "*simulating*" a realistic physical system. Therefore, computer simulations are performed by allowing the particles in a system to move under a certain physical rule. This fundamental aspect of computer–simulation methods is directly related to the advantages of these methods which are essentially the combination of the advantages of both experimental and theoretical physics. A detailed description of these advantages is given in a succeeding section.

The methods of computer simulations became possible when the speed and capacity of computers reached some reasonable standard nearly 40 years ago. Since then, a considerable amount of fruitful results has been obtained. An outstanding success has been attained not only in the detailed numerical analyses of many–particle systems but also in the discovery of concepts and facts which had never been predicted previously either from analytical theories nor from physical intuition.

A remarkable progress in computer technology for the last few decades has made computer simulations all the more powerful for various problems. Today, the computer simulations together with the derivations of numerical results from analytical equations constitute a new branch of methodology for researches in physics. This new branch is classified as "*computational physics*" or "*computer–assisted physics*." As a matter of fact, this new branch of computational physics is evaluated so highly that it is generally accepted as "*a third means for the investigation of physics*" besides experimental physics and theoretical physics.

2. Merits and Extents of Computer Simulations

The status of computer simulations is by no means an expedient subsidiary to the main stream of theoretical studies. It is also inappropriate to regard the techniques of computer simulations as a method of last resort for complicated problems which cannot be handled by analytical theories. In fact, this new branch of research is expected to open a revolutionary era in the history of natural science, and it has indeed brought about a series of breakthrough in the field of physics for the last few decades.

The reason for this success comes from the fact that the investigations in terms of computer simulations have aspects superior to conventional experimental and theoretical studies. Their advantages are clearly manifested by describing the ways by which they overcome the extents and difficulties of experiments and theories. Along this line, the possibilities of computer simulations will be discussed in the present section.

2.1. Comparison with Experimental Physics

The extent of experimental physics is characterized by the following features.

- **a** There exist some materials or situations which are experimentally inaccessible.
- **b** There are some physical properties which cannot be observed experimentally.

c It is rather difficult to specify all the factors which constitute experimental environments.

d There are still quite a few experimental conditions which cannot be realized even by the state–of–the–art apparatus.

On the other hand, the techniques of computer simulations, as stated in what follows, clear all these points, and provide a scheme of much wider scope in the elucidation of fundamental physics.

a* Various kinds of model systems are produced in computers, including those which do not exist in reality or those which are not available at hand. For example, it is possible to construct systems composed of particles interacting via imaginary model potentials for the sake of extracting essential aspects. Another example is the simulations of the situations which are expected to take place in space.

b* A wide variety of physical properties can be measured, ranging from those which are observed in laboratories to those which are far beyond the reach of the latest technology. For example, the full information about the positions and velocities of all particles in a simulation system is obtained, from which any desired microscopic physical quantities are derived such as the mean square displacements and the velocity auto–correlation functions. These quantities can never been measured directly from experiments.

c* Since the input conditions are controlled at our disposal, the cause and effect between input parameters and output data can be investigated without any ambiguity.

d* Some extreme conditions which are not achieved in laboratories can be realized. To name but a few, very high pressures, very low temperatures and very rapid quench rates are among them.

2.2. Comparison with Theoretical Physics

The scope of theoretical physics is characterized by the following points.

a Although some simplified (often too simplified) models can be solved rigorously, the physical relevance of the solutions is not always obvious.

b For complicated models, problems must be solved by making use of some approximations, but the validity of approximations is not necessarily guaranteed from physical point of view.

The approaches by means of computer simulations are free from these limitations of theoretical physics as explained in the following.

a* Model systems for computer simulations do not have to be extremely simplified, since the complexity of a certain degree does not introduce serious problems here. As a result, it becomes possible to study models much closer to realistic materials than the models tractable by analytical theories. In fact, computational physics can bridge a gap between experimental and theoretical physics.

b* Computer simulations lead to true solutions without invoking an aid of approximations. Accordingly, it is not necessary to worry about the validity of approximations, which is often the case in theoretical approaches. As a consequence, the results obtained from computer simulations not only give straightforward answers but also serve as the standards, with which the outcomes from approximate theories should be compared and judged.

2.3. Extent of Computer Simulations

There are a couple of points about computer simulations which we have to bear in mind.

a **Finite–size effects**

Except for problems concerning microclusters or fine particles, most systems of interest are in the thermodynamic limit where both the number of constituting particles and the volume of the system are infinite, while computer simulations deal with systems of finite sizes. For the purpose of minimizing the finite–size effects, appropriate boundary conditions are applied such as the periodic boundary conditions. But, we must remember that there are some physical quantities which are rather sensitive to boundary conditions.

b **Finite–observation–time effects**

In computer simulations, a desired physical quantity is evaluated by taking an average over timesteps or over different MD runs starting from different initial conditions. The underlying concept for taking averages is based upon the fact that the experimentally measured value for that physical quantity is actually the average over the time which elapses during the measurement. A point to make, however, is that the simulation time is much shorter than the duration needed for the experimental measurement, and inevitably there appear the finite–observation–time effects in the results of computer simulations.

Although these effects might be reduced to a certain extent when the performance of computers is largely improved, it is important to try and understand the ways by which these effects are reflected in the physical quantities of simulated systems.

3. Molecular Dynamics Simulations

The basic scenario of the MD simulations was already written in 1950's by the forerunners of this field. The scenario is composed of the following four steps.

1 To set up the classical equations of motion for N particles in a system according to classical mechanics.

The Newtonian equations of motion are expressed by the following well–known form of N coupled differential equations.

$$M_n \ddot{R}_n = F_n(\{R_n\}) \quad (n = 1, ..., N).$$

Here, M_n denotes the mass of nth particle, $\{R_n\}$ the position vectors $R_1, .., R_N$ of N particles, and $F_n(\{R_n\})$ the force on nth particle from all other particles in the system. The two dots on R indicate the second derivative with respect to time, thus $\ddot{R}_n$ being identical to the acceleration of nth particle.

2 To replace the differential equations of motion by the corresponding difference equations.

3 To trace the time developments of motions of these particles by solving the difference equations thus obtained.

As can be readily seen from these steps of the MD simulations, the motions of the particles are determined by the rules of classical mechanics, and consequently, the time dependences of the particle motions derived therefrom are, when compared to other simulation methods, more closely grounded on the physical rules.

4 To calculate various physical properties of the system by making use of information about the positions and velocities of all N particles.

This original scenario of the very early days in the history of the MD simulations was almost perfect so that nothing drastically new has been added until now.

In order to proceed actual MD steps on the basis of the above–described Newtonian equation, it is necessary to know the exact expressions of the forces acting on each particle in the system, that is to say, the right–hand side of the equation, $F_n(\{R_n\}) \quad (n = 1, ..., N)$. As usual, the forces are derived from the potential energy $E(\{R_n\})$ of the system by the relation;

$$F_n(\{R_n\}) = -\nabla_n E(\{R_n\}) \quad (n = 1, ..., N)$$

where the potential energy $E(\{R_n\})$ is generally written as the sum of interatomic potentials.

When we say that we choose a model for MD simulations as stated in section 2.2, it practically means that we choose some explicit forms of jnteratomic potentials. Since there is a wide variety of choices for interatomic potentials, a rather comprehensive study of many classes of systems become possible by the MD simulations.

As a most simplified model, the total potential of a system is often written as a sum of pair interatomic potentials for all pairs of particles in the system. So far, quite a few different types of pair potentials have been investigated. They include hard core potentials and soft core potentials which are studied for the sake of clarifying the basic properties of many–particle systems, the Lennard–Jones potentials which are proved to be appropriate for closed–shell materials, the Stillinger–Weber potentials which are modified to become useful for ionic systems, and potentials which are proposed for simple metals such as alkaline metals. Many attempts of MD simulations have been carried out by adapting these model potentials and a considerable amount of information has been accumulated by now.

4. Fashionable Trend of Molecular Dynamics Simulations

In the history of the MD simulations, the early days of 1950's and 1960's are the dawn when the fundamental basis were implanted as described in the preceding section. Even some important results of interest were already obtained in this period. These early days were followed by the era of growth in 1970's when the idea of the MD techniques was applied to various problems.

Activated by the rapid progress of computers, the years of 1980's were the era of expansion of the MD studies. The characteristic of this era is that the traditional MD methods were reconsidered from several viewpoints, which led to new attempts at expanding the extent of the MD techniques. A few examples are mentioned in what follows.

1. An attempt to carry out the MD simulations of large systems. Although the increase in the capacity and speed of computers makes it possible to handle the sizes larger than those treated previously, it would be dishonourable for physicists to depend totally upon the progress of computers. It is required to establish efficient techniques which enable us to deal with systems larger than those which the progress of computers allows us to.
2. An attempt to realize ensembles of desired types. The physical properties of a system constructed by MD simulations are derived in the forms of averages. In order for the quantities thus obtained to be compared with experimental data, it must be guaranteed that the ensemble produced by the MD simulations closely simulates the experimental conditions.
3. An attempt to study systems in non–equilibrium states, which are expected to provide the kind of information concerning the problems

of non–equilibrium systems that is inaccessible through other methods.

4. An attempt to perform the MD simulations from the first principles in the sense that the forms of interatomic potentials are determined self–consistently without making assumptions. It is crucial, however, that the computation of interatomic potentials must be efficient enough so that the whole simulations are within the reach of today's computers. The establishment of the first–principle or ab–initio MD methods is good news for systems such as covalent materials in which the assumptions of pair potentials fails.

As fully described in the papers in these proceedings, the mentioned new attempts have been developed successfully in 1980's. They are now increasing the possibilities of the MD business, and we expect the present decade 1990's to be the era of maturity of the MD methods.

5. Outline of This Volume

It is rather difficult to classify categorically the papers presented at the Symposium, because each paper has several aspects. The division here into Parts is one possible solution by *"optimization."*

In Part 1, the introductory descriptions of the MD simulations are given. In particular, a detailed explanation of the new trend in 1980's is made by Nosé.

The area in which the MD methods get credit includes the problems of disordered systems such as liquids, glasses and amorphous materials. Papers dealing with disordered systems are given in Parts II to IV. To be specific, Part II is mainly concerned with simple systems and Part III is related to more complex systems while, in Part IV, liquids are treated from a viewpoint of somewhat larger scale.

The ab–initio MD methods are discussed at length in Part V. Some recent attempts for nonequilibrium MD techniques are described in Part VI, while the problems of interface and ionic conductors are presented in Part VII.

Acknowledgement

The author is grateful to Mr. T. Yamaguchi for his help in the technical arrangements for the publication of these proceedings.

The Development of Molecular Dynamics Simulations in the 1980s

S. Nosé

Department of Physics, Faculty of Science and Technology, Keio University, 3-14-1 Hiyoshi, Kohokuku, Yokohama 223, Japan

Abstract.
The 1980s was a decade of revolutionary change in molecular dynamics simulations. Two key factors were the appearance of new methods and a remarkable increase in computer performance. Three new methods (the simulations at constant temperature and/or pressure, the nonequilibrium molecular dynamics and the ab initio molecular dynamics methods) have significantly extended the range of the computer simulations. Large scale simulations were also a trend in the 1980s. They are applied to fluid dynamical problems, to slow relaxation in a supercooled liquid, and to the engineering problems in material sciences.

1. Introduction

The molecular dynamics method is one of the computer simulation methods employed to study the properties of a many-particle system. When particles gather together, they show properties different from those of a single particle. They can be in several different phases, e.g. crystals, liquids, or liquid crystals. Each phase exhibits its characteristic structure and properties, and a phase may change into another phase when the temperature or the pressure is changed. The phase transition is one of the typical phenomena appearing in a many-particle system.

The theoretical study of an interacting many-particle system is a very difficult task. The system is generally too complicated to be studied analytically by first principles. Thus, in almost all cases, it is necessary to introduce approximations of some kind to solve the problem. Sometimes it is very difficult to distinguish whether the result is a genuine one or an artifact arising from approximations.

The invention of the electronic computer opened up a new possibility of relating the macroscopic properties of matter to the microscopic information on atoms and molecules. The main idea of computer simulation is as follows. The configurations of a many-particle system are reproduced by numerical computations, and thermodynamical properties are calculated by averaging certain quantities over configurations thus obtained. In the molecular dynamics method, the equations of motion of classical mechanics are integrated numerically and the movement of each particle is followed. In the Monte Carlo method, another typical simulation method, a stochastic process is employed to change the configurations.

The development of molecular dynamics simulation is reviewed by putting emphasis on the methodological aspect. The two main factors in the development of computer simulation in the 1980s were innovation in simulation methods and a rapid increase in computer performance.

In section 2, we review the development of the methodology in molecular dynamics simulation until 1980.

In section 3, three revolutionary changes in molecular dynamics methods proposed in the 1980s are reviewed. They are simulation in ensembles other than the microcanonical ensemble, the nonequilibrium molecular dynamics, and the ab initio molecular dynamics, which takes into consideration the behavior of the electrons in a system. These new methods have really extended the range of fields that we can study by computer simulations.

In section 4, several examples of the new trends are exhibited. Large scale simulations are promoted by the remarkable improvement in computer performance.

2. Early days

The study of the properties of a liquid state has been a challenging problem for computer simulations. Contrary to gas and solid states, there is no simple model to describe well the nature of a liquid state. In a gas state, a noninteracting ideal gas is a good model, and we can learn much about the nature of a solid from a perfect crystal model. Therefore, it was not surprising that many of the computer simulations have been devoted to elucidating the nature of a liquid.

In the 1950s, two basic computer simulation methods, the Monte Carlo [1] and molecular dynamics [2] methods, were introduced. Already at this early stage, many of the basic techniques of computer simulations (periodic boundary conditions, the truncation of the interaction potential, and its correction) were employed.

Alder and Wainwright [2,3] applied the molecular dynamics method to the study of hard-sphere systems. This is generally considered as the first molecular dynamics simulation. The discovery that a fluid-solid phase change exists even in a hard-sphere system without an attractive force was a surprise. The simulations in these early days were pioneering works in every respect, they had a great impact by the discovery of unexpected results, and confirmed the importance of the numerical approach for the many-body problem.

A system with a continuous interacting potential was investigated by Vineyard et al. [4] in a study of the radiation damage of a crystal and by Rahman [5] in a study of a Lennard-Jones liquid. Vineyard's work was the first computer simulation application to the material sciences. Rahman's work was also a milestone; he investigated many dynamical properties in liquids.

Verlet [6] determined a phase diagram of a Lennard-Jones system. In this study, he introduced the so-called Verlet algorithm (the most popular numerical integration method in the molecular dynamics simulations).

The simulation of liquid water by Rahman and Stillinger [7] marked the start of the 1970s. This also indicated a change of the trend in simulation. The target of study changed from a simple model system to a realistic one. Rahman and Stillinger used an empirical potential for their study of liquid water, but afterwards the interactions between molecules were often determined by the aid of molecular orbital calculations. A typical example is the work by Clementi's group on water-water interactions [8,9].

Another typical numerical integration method, a predictor-corrector algorithm (Nordsieck or Gear algorithm [10,11]) is introduced in Rahman and Stillinger's work.

The treatment of the rotation of a rigid body becomes important in the study of molecular systems. The equations of motion for the molecular rotation expressed by using Euler angles have singular points. This is an undesirable feature for numerical integration. Singularity-free algorithms for molecular rotation were introduced by Evans [12] and by Berendsen et al. [13]. Evans rewrites the Euler equations for the molecular rotation in terms of quarternions. In this form, the singularities disappear from the equations of motion and it has also the merit that the calculation of the trigonometric functions is not necessary. In the method by Berendsen et al., a molecule is considered to consist of a set of atoms. The equations of motion for an atomic system are solved under the imposition of constraints to fix atomic arrangements in a molecule.

The calculation of the free energy evoked much interest in the 1970s. The most promising method is an umbrella sampling method [14] based on a weighted average over two different states. A recent remarkable extension in this line is the method proposed by Frenkel and Ladd [15]. They make possible the direct comparison of the free energy between different crystal structures by gradually replacing the interatomic interactions by an Einstein lattice vibration model.

3. New methods appeared in the 1980s

The 1980s was a decade of revolution for the molecular dynamics method. Many new algorithms were presented. They will be categorized into the following three groups: (1) simulations in ensembles different from the traditional microcanonical ensemble. This means simulations at constant temperature and pressure. (2) Nonequilibrium molecular dynamics method, (3) Ab initio molecular dynamics method known as the Car-Parrinello method. These methods are related intrinsically and inspired the development of the methods in different categories.

A common feature of new methods is that they modify the equations of motion in classical mechanics. This modification allowed the adaptation of the computer simulations to various situations.

3.1 Simulations at constant temperature and pressure

Most simulation methods in an ensemble different from the traditional microcanonical ensemble originate from the work by Andersen [16]. He introduced a new idea to add a new degree of freedom to a physical system. In his formulation of the constant pressure simulation method, the volume becomes a dynamical variable and the change of the volume is driven by the difference between an internal pressure and an external pressure. The internal pressure is expressed as the average of the kinetic energy and the virial and reflects the microscopic information of the particle motion and the interaction between particles. On the other hand, the external pressure is a controllable parameter of the simulation. A feedback mechanism works to keep the pressure at a constant value. If the internal pressure is larger than the external pressure, the volume expands and the internal pressure decreases. When the internal pressure becomes smaller than the external pressure, the direction of the volume change is inverted and the volume shrinks. In this way, the internal pressure is controlled to fluctuate around the external pressure and the average of the pressure is kept equal to a constant. The value of the volume is determined from the balance of the microscopically determined internal pressure and the external pressure. This formulation is a natural way to realize a constant pressure condition.

Parrinello and Rahman [17,18] extended Andersen's constant pressure method to allow a change of the shape of a basic molecular dynamics unit cell. With this extension, we can directly simulate the structure change in a solid. This extension of the constant pressure method is formulated so that particles choose a stable structure by themselves. This is a powerful tool to investigate the phase transitions in crystals.

Nosé [19,20] introduced a constant temperature method by adding a degree of freedom to a physical system. The new degree of freedom is interpreted to correspond to a heat bath. Hoover [21] reformulated the equations proposed by Nose to simple ones. This reformulation clarified a feedback mechanism to keep the temperature at a constant value. A term like a friction force is added to control the total kinetic energy. This approach is quite general and can be extended to allow separate temperature controls for different degrees of freedom [22]. The temperature control in a system without a kinetic energy term is discussed in recent work by Kusnezov and Bulgac [23]. A classical spin system is an example of such a system.

A constant pressure and temperature condition is also realized by the constraint methods [24,25] and the stochastic methods [16,26]. A Gaussian thermostat method [24] solves the equations of motion of classical mechanics under the constraint of a constant kinetic energy. This method is derived from the study of nonequilibrium molecular dynamics simulation. From the analysis of this Gaussian thermostat method, the nature of the velocity scaling algorithm proposed by Woodcock [25], sometimes called an ad hoc velocity scaling method, is clarified. It was proposed in 1971, but for a long time it was not clear whether this method really produced a canonical distribution or not. Now, the velocity scaling algorithm is considered as an approximate algorithm for the Gaussian thermostat method.

3.2 Nonequilibrium molecular dynamics

A driving motivation in the nonequilibrium molecular dynamics method was the calculation of the transport properties of a fluid (e.g. the diffusion coefficient, viscosity and thermal conductivity). There are two ways to approach this problem. One is the investigation of the response to the externally set nonequilibrium condition. This is in principle the same approach as in laboratory experiments. Another is to employ the fluctuation dissipation theorem of nonequilibrium statistical mechanics. The correlation of the fluctuation in equilibrium is related to the response to the perturbations added to a system. The velocity autocorrelation function is related to the diffusion coefficient, which is a response to a perturbation which causes a density gradient. A problem in the latter approach is that simulations over a large number of time steps are necessary to obtain a reliable result.

If we realize a nonequilibrium situation by imposing externally set boundary conditions, the particles in a surface region are affected by the boundary, and behave differently from those in a bulk region. A weak point of the simulation, the smallness of the system size, becomes apparent in this type of simulation.

A new approach introduced to overcome this difficulty is the nonequilibrium molecular dynamics method in a narrow sense [27]. The equations of motion are modified to reproduce the nonequilibrium external condition. The form of the equation is selected to conform with the periodic boundary condition and to reproduce a correct response in accord with the fluctuation dissipation theorem to a perturbation added to a system [28]. For example, the viscosity η is measured as a ratio of the xy component of the stress tensor P_{xy} and the velocity gradient $\gamma = \partial v_x / \partial y$ of a shear flow, $\eta = -P_{xy}/\gamma$. The shear stress is calculated for finite γ values, and the extrapolation to $\gamma = 0$ gives the viscosity in equilibrium.

In the application to the finite shear rate problem, a new type of ordered flow, the string phase, was discovered in the study of a planar Couette flow by Erpenbeck in 1984 [29]. The particles align in a string and the strings form a triangular lattice in a plane perpendicular to the flow. The reliability of the simulation method to realize the string phase was questioned recently [30,31], but this structure is a remarkable example of the unexpected discovery of new phenomena by computer simulations.

3.3 Ab initio molecular dynamics method

The third innovation of the molecular dynamics method in the 1980s derived from the calculation of electronic properties in condensed matter physics. Nowadays, the electronic structure of a solid is fairly well understood from the calculation based on the density functional theory. It requires self-consistent calculations of the wave functions of the electrons in a fixed atomic configuration. The atomic configuration of atoms is changed in search of a stable structure. The atoms are moved by the Hellmann-Feynman force, which is calculated from the wave function thus obtained. The diagonalization of a large matrix and the optimization of the atomic configurations were two time-consuming factors in this process.

Car and Parrinello [32] proposed a new algorithm which can reduce the computational efforts significantly in the process to obtain an optimized stable structure. They coupled the calculation of the electronic properties in the density functional theory and the atomic motion by the molecular dynamics simulation. A classical mechanical kinetic energy term for the wave functions is added to describe the time evolution of the wave function in a classical mechanical sense. The diagonalization of the wave equations is carried out by reducing the kinetic energy for the wave functions. If it reaches a minimum energy state, the most stable atomic configuration and the correct wave functions will be obtained. It is not necessary to prepare a large matrix of size M^2 (M is the number of basis functions) which is required in a usual matrix diagonalization, and in the Car-Parrinello method only a region of size NM (N is the number of the wave functions) is necessary (generally $N \ll M$). The optimization of the wave function and the atomic structure is carried out simultaneously. These two features are the major factors accelerating the calculation in the Car-Parrinello method.

We have considered the Car-Parrinello method from the viewpoint of electronic structure calculations. In this field, the method is considered as a new algorithm which will reduce the computing efforts drastically.

From the viewpoint of the molecular dynamics simulations, this method is considered as a new method which gives the possibility of carrying out a simulation on the basis of quantum mechanics. The interacting forces between atoms are generally determined from calculation of the electronic properties of the system. The electronic properties are the determining factor in a system with covalent bondings or in a metal, and in this case it is important to take into consideration the interrelation between the electronic properties and the atomic configuration. If we use the traditional method to diagonalize large matrices for simulations in such a system, we can only apply it to a system of at most the order of ten atoms at present. If we employ the Car-Parrinello method, the calculation in a system of one hundred atoms is possible.

4. Large scale simulations

As well as the innovation in simulation methods, a remarkable increase in computer performance is another factor in the development of computer simulations. In the early days, simulation was limited to a simple system with a small number of particles. Through this period, it was confirmed that we can reproduce fairly well the macroscopic properties for a single phase in equilibrium in a system of particles of the order of hundreds.

Nowadays, we can apply the increasing computing power to more complex problems. The records in system size and in time steps are renewed frequently. Even a simulation with more than one million particles has already been reported by a Lawrence Livermore group [33]. Promising applications for large scale simulations are inhomogeneous and time-dependent nonequilibrium problems.

A good example of simulation in a large system is a challenge to the fluid dynamical problem by particle simulations. A two-dimensional flow passing a circular obstacle was studied by Rapaport and Clementi [34]. This is a well studied problem of fluid dynamics. The behavior of the flow is characterized by the Reynolds number $R=dv/\eta$; where d is a characteristic size of the obstacle, v is the velocity, and η is the viscosity. As the Reynolds number increases, the flow changes its character from a laminar flow to a flow accompanying symmetrical counterrotating eddies behind the obstacle, to asymmetrical oscillating wakes, to the famous von Karman's vortex street, and to a turbulent flow. In the simulation by Rapaport and Clementi, the fluid is not a continuous medium, but consists of many particles. The macroscopic flow thus obtained from a particle simulation really exhibits the same behavior as in fluid dynamics. A symmetrical eddy changes to time-dependent asymmetrical oscillating wakes. A series of eddies continues like the von Karman's vortex street. Simulations in a system with an enormous number of particles will give new insight and this possibility is demonstrated in this work.

The dynamical relaxation in a supercooled liquid becomes very slow as the temperature approaches the glass transition point. Therefore, simulation with a large number of time steps is necessary to study this problem. The change in the dynamical character of a supercooled liquid has become obvious from studies in recent years [35]. An exponential decay law $\exp[-At]$ in the incoherent part of the density autocorrelation function changes into a stretched exponential decay $\exp[-At^{\beta}]$ near the glass transition, where $\beta<1$.

A glass state is obtained by quenching a supercooled liquid. This state is in nonequilibrium and the structure and properties depend on the history of the quenching. The study of a glass has been one of the major interests in computer simulations.

A system consisting of large molecules is also a challenging field. Liquid crystals, biopolymers and a system with surfactants are some examples. The computer simulation demonstrates its power in seemingly intractable complex problems.

Finally, we would like to comment on the application of computer simulation to problems in material sciences. The structure around a defect, or a grain boundary, and fractures in a crystal are some of the engineering problems we can now study by computer simulation. A large scale simulation is often necessary in this type of problem. A simulation with more than one million particles [33] is applied to an indentation problem. In this simulation, zigzag cracks penetrate into a crystal hit by an object.

5. Conclusion

The developments in molecular dynamics simulations in the 1980s are reviewed. The proposals of new simulation methods and large scale simulations promoted by a remarkable increase in computer performance are two major features which characterize the progress in this period. These innovations extend considerably the application of computer simulation from a simple homogeneous system in equilibrium to inhomogeneous, nonequilibrium, and complex problems.

References

[1] N.Metropolis, A.W.Rosenbluth, M.N.Rosenbluth, A.H.Teller, and E.Teller, J.Chem.Phys. 21, 1087 (1953).
[2] B.J.Alder and T.E.Wainwright, J.Chem.Phys. 27, 1208 (1957).
[3] B.J.Alder and T.E.Wainwright, Phys.Rev. 127, 359 (1962).

[4] J.B.Gibson, A.N.Goland, M.Milgram, and G.H.Vineyard, Phys.Rev. **120**, 1229 (1960).
[5] A.Rahman, Phys.Rev. **136A**, 405 (1964).
[6] L.Verlet, Phys.Rev. **159**, 98 (1967).
[7] A.Rahman and F.H.Stillinger, J.Chem.Phys. **55**, 3336 (1971).
[8] H.Popkie, H.Kistenmacher, and E.Clementi, J.Chem.Phys. **59**, 1325 (1973).
[9] O.Matsuoka, E.Clementi, and M.Yoshimine, J.Chem.Phys. **64**, 1351 (1976).
[10] A.Nordsieck, Math.Comput. **16**, 22 (1962).
[11] G.W.Gear, Numerical Initial Value Problems in Ordinary Differential Equations (Prentice Hall,Englewood Cliffs, NJ) 1971, Chap. 9.
[12] D.J.Evans, Mol.Phys. **34**, 317 (1977).
[13] J.P.Ryckaert, G.Ciccotti, and H.J.C.Berendsen, J.Comput.Phys. **23**, 327 (1977).
[14] G.M.Torrie and J.P.Valleau, J.Comput.Phys. **23**, 187 (1977).
[15] D.Frenkel and A.J.C.Ladd, J.Chem.Phys. **81**, 3188 (1984).
[16] H.C.Andersen, J.Chem.Phys. **72**, 2384 (1980).
[17] M.Parrinello and A.Rahman, Phys.Rev.Lett. **45**, 1196 (1980).
[18] M.Parrinello and A.Rahman, J.Appl.Phys. **52**, 7182 (1981).
[19] S.Nosé, Mol.Phys. **52**, 255 (1984).
[20] S.Nosé, J.Chem.Phys. **81**, 511 (1984).
[21] W.G.Hoover, Phys.Rev. **A31**, 1695 (1985).
[22] S.Nosé, Mol.Phys. **57**, 187 (1986).
[23] D.Kusnezov and A.Bulgac, private communication.
[24] D.J.Evans, W.G.Hoover, B.H.Failor, B.Moran, and A.J.C.Ladd, Phys.Rev. **A28**, 1016 (1983).
[25] L.V.Woodcock, Chem.Phys.Lett. **10**, 257 (1971).
[26] T.Schneider and E.Stoll, Phys.Rev. **B17**, 1302 (1978).
[27] D.J.Evans and G.P.Morriss, Comput.Phys.Rep. **1**, 297 (1984).
[28] D.J.Evans and G.P.Morriss, Phys.Rev. **A30**, 1528 (1984).
[29] J.J.Erpenbeck, Phys.Rev.Lett. **52**, 1333 (1984).
[30] D.J.Evans and G.P.Morriss, Phys.Rev.Lett. **56**, 2173 (1986).
[31] W.Loose and S.Hess, Rheol.Acta **28**, 91 (1989).
[32] R.Car and M.Parrinello, Phys.Rev.Lett. **55**, 2471 (1985).
[33] W.G.Hoover, A.J.De Groot, C.G.Hoover, I.F.Stowers, T.Kawai, B.L.Holian, T.Boku, S.Ihara, and J.Belak, Phys.Rev. **A42**, in press (1990).
[34] D.C.Rapaport and E.Clementi, Phys.Rev.Lett. **57**, 695 (1986).
[35] Y.Hiwatari and H.Miyagawa, J.Non-Cryst. Solids **117/118**, 862 (1990)

Part II

Supercooled Liquid and Glass Transition

Molecular Dynamics Studies of Diffusion in Liquids

*J.-L. Barrat**

Department of Nuclear and Chemical Engineering, University of California, Santa Barbara, CA 93106, USA
*Permanent address: Laboratoire de Physique, ENSL,
46 Allée d'Italie, F-69007 Lyon, France

Abstract : Molecular dynamics simulations are used to investigate diffusion phenomenon in simple liquids. Three examples are discussed: translational and rotational diffusion in supercooled liquids, and tracer diffusion in a normal liquid. In all three cases, the simulations give access to quantities which are not easily measurable experimentally, and reveal interesting crossover phenomena.

1. Introduction

Among the often listed interest of the Molecular Dynamics (MD) method is the fact that it gives access to numerous properties which are not experimentally measurable. Strictly speaking, diffusion does not belong to this category, since it can be characterized using various experimental techniques, such as incoherent neutron scattering, NMR, or macroscopic tracer diffusion. MD simulations, however, provide a very direct information on quantities which would require a rather cumbersome deconvolution procedure were they to be extracted from experimental data. Moreover, the observation of particles trajectories can often give a direct insight into the nature of the diffusion process at the microscopic level. Finally, the study of diffusion is made easier in MD simulations by the fact that it is a single particle property. Very good statistics are thus are therefore achieved by averaging over the few hundreds particles present in the sample.

In this paper, I shall present three examples of diffusion studies in liquids using the MD method. I shall be mostly concerned with translational diffusion. The relevant quantities for the description of translational diffusion are:

-The diffusion constant D, defined as the long-time slope of the mean-squared displacement. D can be obtained either from this definition (Einstein formula)

$$D = \lim_{t \to \infty} \frac{1}{6t} < (\vec{r}(t) - \vec{r}(0))^2 > \tag{1}$$

or from the equivalent Kubo formula

$$D = \frac{1}{3} \int_0^\infty < \vec{v}(t).\vec{v}(0) > dt\,. \tag{2}$$

In a liquid, a typical order of magnitude for D is $10^{-5} cm^2 s^{-1}$. Note that when D is small, (1) is more accurate than (2), since in this case a cancellation between the positive (short-time) and negative (long-time) parts of the velocity autocorrelation function takes place.

-The incoherent scattering function $F_s(k,t)$, defined as:

$$F_s(k,t) = N^{-1} \sum_{i=1}^{N} exp(i\vec{k}.(\vec{r_i}(t) - \vec{r_i}(0)))\ . \tag{3}$$

For small wavevectors k, $F_s(k,t)$ in a liquid behaves like $exp(-k^2 Dt)$. $F_s(k,t)$ (or rather its Fourier transform $S_s(k,\omega)$) is measured in incoherent neutron scattering experiments.

-The van Hove correlation function, $G_s(r,t)$, is the spatial Fourier transform of $F_s(k,t)$. Its definition is:

$$G_s(r,t) = N^{-1} \sum_{i=1}^{N} < \delta(\vec{r_i}(t) - \vec{r_i}(0) - \vec{r}) > \ . \tag{4}$$

The quantity $P(r,t) = 4\pi r^2 G_s(r,t)$ gives the probability density for finding a particle at a distance r from the origin at time t knowing it was at the origin at time 0. At long times, $G_s(r,t)$ in a liquid is given by the equations of macroscopic hydrodynamics and reads:

$$G_s(r,t) = \frac{1}{(4\pi Dt)^{3/2}} exp(-\frac{r^2}{4Dt})\ . \tag{5}$$

2. Diffusion in a supercooled binary mixture

As a liquid is cooled below its freezing point, the diffusion constant decreases dramatically, while the viscosity correspondingly increases. This increase gives rise to the so-called calorimetric glass transition, observed when the relaxation times of the system become larger than the experimental time scales. In what follows, I shall describe a different phenomenon which was observed in a soft sphere mixture at a temperature slightly above the "computer glass transition" for this system [1,2]. The system under consideration is made up of spherical particles interacting through a short-range repulsive potential, $v(r) = \epsilon(\sigma/r)^{12}$. The use of a mixture is justified by the need to avoid crystalline nucleation, but is otherwise irrelevant. For this system, the temperature T and density n are not independent variables, so that cooling and compression are equivalent. (The reader is referred to references [3,4] for more details on thermodynamic properties)

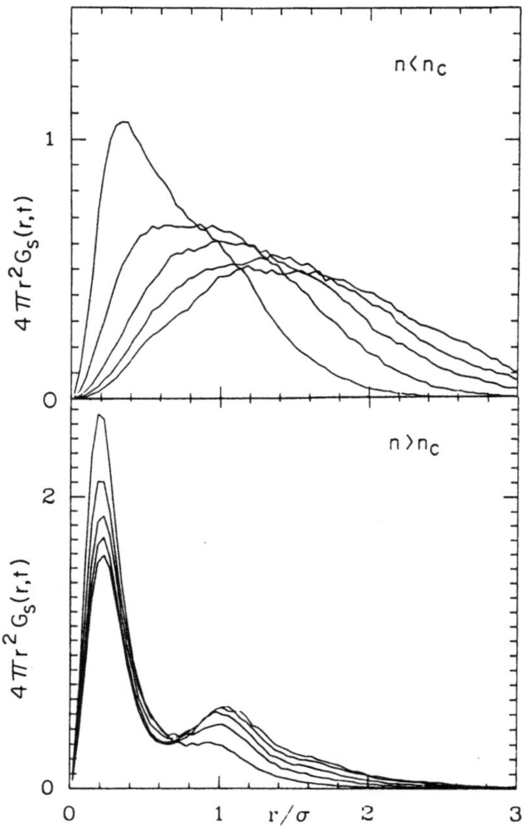

Top: P(r,t) versus r curves for different times in a soft sphere system at a density $n = 1.43$, slightly below the crossover density $n_c = 1.44$. From right to left, the times are $t/\tau = 100, 200, 300, 400, 500$, where $\tau = (m\sigma^2/\epsilon)^{1/2}$. Bottom: same curves, but now at a slightly higher density, $n = 1.45$. From reference [2].

Figure 1 illustrates the behaviour of the function $P(r,t)$ for two densities respectively below and above a crossover density n_c. For $n < n_c$, $P(r,t)$ as a function of r presents a single maximum. This maximum shifts with increasing times towards larger values of r, and the curve eventually reaches its asymptotic form, equation (4). A completely different behaviour is observed for $n > n_c$. $P(r,t)$ is now a bimodal distribution. The height of the seconf peak increases with time at the expense of the first one, but the peaks positions hardly change. This is exactly what would be expected for jump diffusion in a solid. In a perfect solid, $G_s(r,t)$ would behave for long times like:

$$G_s(r,t) = (\frac{\alpha}{\pi})^{3/2} exp(-\alpha r^2) \tag{6}$$

and P(r,t) would thus remain single-peaked. If jump diffusion takes place, one expects a second peak to appear at the position corresponding to the next nearest neighbour distance. In principle, the fit of $G_s(r,t)$ to a jump diffusion model could allow a determination of jumping rates. This is unfortunately, quite difficult to realize in practice: while the results obtained for $n < n_c$ are perfectly reproducible, the shape of $G_s(r,t)$ becomes sensitive to system history for $n > n_c$. The bimodal character is preserved, but the amplitude of the peaks varies from sample to sample. One aspect of the jump diffusion process which is not apparent in the $P(r,t)$ curves, but can be inferred from a study of particles trajectories, is that a jump event usually involves three to four neighbouring particles [5].

There is some evidence that this crossover from "liquid-like", or "hydrodynamic" diffusion to a "solid-like" jump diffusion is a rather generic phenomenon, at least in the so-called "fragile" liquids -i.e. liquids made up of particles with simple, close to spherically symetric interactions. Firstly, the existence of such a phenomenon was postulated by Goldstein [6] more than twenty years ago, on the basis of several order of magnitude estimates. He concluded that when typical relaxation times become about $10^{-9}s$, the system trajectory in phase space comes close enough to its potential energy surface that the dynamics becomes dominated by activated jumps between minima on this surface. This time scale corresponds to the one for which the crossover is observed in MD simulations. The relaxation time can be estimated from the ratio η/G_∞, where η is the shear viscosity ($\eta \sim 10 - 100$ Poise at $n \sim n_c$) and G_∞ the high frequency shear modulus.

A second, more quantitative indication comes from the work of Rössler [7]. From a Stokes-Einstein analysis of diffusion and NMR data, this author shows that a crossover temperature exists, below which Stokes-Einstein type relations dramatically fail, and would actually strongly underestimate the diffusivity. The viscosities at this crossover temperature are compatible with those observed at n_c in the MD simulations. The fact that a Stokes-Einstein relation underestimates D is also easily understood from the simulation results: in a solid, one can have a very large (in practice, infinite) viscosity with a non-zero diffusion constant due to vacancies.

Finally, it is interesting that the mode-coupling theories [8,9,10] of supercooled liquids, which are expected to describe a liquid-like behaviour quite well, give good results for the viscosity and relaxation properties for $\eta < 10 - 100$ Poise, and relaxation times $\tau < 10^{-9} - 10^{-8}s$. These theories then predict a power-law divergence of the viscosity which is not experimentally observed (Indeed, one rather observes a crossover to Arrhenius behaviour). Thus it is tempting to conclude that the point at which mode-coupling theories break down is the point at which both experiments and simulations detect a crossover from liquid-like to solid-like in the diffusion mechanism.

3. Rotation of nitrate ions in a nitrate melt

A phenomenon quite similar to the one described above for translational diffusion was found to occur in the rotational motion of NO_3^- ions in a supercooled nitrate melt, $[Ca\ (NO_3)_2]_{0.4}[K\ (NO_3)]_{0.6}$. This fragile glass-forming ionic melt has been widely studied experimentally, and the simulation work [11] was intended to supplement the data from neutron scattering experiments [12]. One main conclusion of the simulation work was that translation and rotation are strongly coupled. This can be demonstrated by studying the temperature dependence of the relaxation times τ_D for $F_s(k,t)$ and τ_i for the rotational correlation $C_i(t)$ functions defined as:

$$C_i(t) = < P_i(\vec{u}(t).\vec{u}(0)) > \quad (7)$$

where $\vec{u}(t)$ is a unit vector perpendicular to the NO_3^- plane and P_i is the Legendre polynomial of order i. Over the temperature range studied, τ_D and τ_2 remain proportional to each other. An interesting decoupling is however observed at a temperature $T_0 \simeq 400K$. Above T_0, τ_2 and τ_1 have similar values, but below T_0 τ_1 becomes much smaller than τ_2. The reason for this decoupling can be seen in the trajectories presented in figure 2. Below T_0, the only motion consists in jumps of amplitude 180 degrees, which contribute to the decay of C_1 but not to that of C_2. At higher temperature, the motion is close to rotationnal diffusion, and $\tau_1 \simeq \tau_2$.

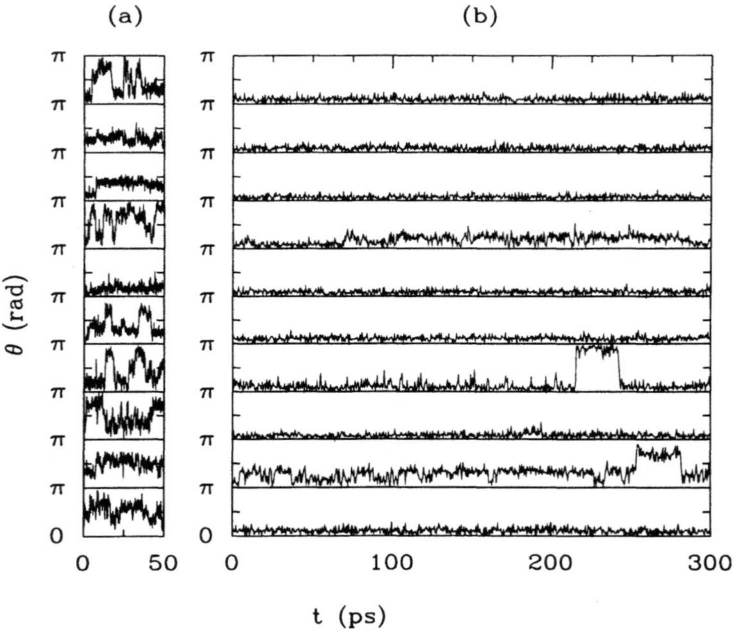

Rotation of nitrate ions in $[Ca\ (NO_3)_2]_{0.4}[K\ (NO_3)]_{0.6}$ as a function of time, at temperatures above (left, $T = 700K$) and below (right, $T = 350K$) T_c. From reference [11].

Once again, it is interesting to note that this decoupling phenomenon is observed at a temperature close to the one identified in neutron scattering work as the "critical temperature" predicted by mode-coupling theories. No jump diffusion was observed in the simulation, likely because the corresponding time scale is too long - a fact that makes this system a good benchmark for testing mode-coupling theories. However, conductivity measurments seem to indicate a crossover in the diffusion mechanism at a slightly lower temperature [13].

4. Tracer diffusion in a Lennard-Jones liquid

The last topic I want to briefly address is not directly related to those discussed above, but still bears some relationship to them. The problem under consideration is the diffusion of a small tracer particle (size σ_2) diluted in solvent of bigger (size σ_1) molecules. While diffusivity of big particles, or even self-diffusion, is well accounted for by the Stokes-Einstein relation,($D \simeq k_B T/\eta\sigma_2$, η beeing the solvent viscosity), the a priori conditions for the validity of this relation are not fulfilled when $\sigma_2 < \sigma_1$. Experimental studies [14] conclude to the existence of a correlation between tracer diffusivity and solvent viscosity of the form $D \simeq \eta^{-\alpha}$, $\alpha \simeq 2/3$. It is not clear, however, whether the solvent viscosity is really the only relevant parameter. MD simulations,on the other hand, allow independent and continuous variations of the different parameters such as tracer mass and size or solvent-solute interaction potential. Here, I shall only present results obtained from variable tracer sizes, the tracer mass beeing kept constant (and equal to the solvent atom mass). The interaction potentials are of the Lennard-Jones 6-12 type, $v_{ij}(r) = \epsilon((\sigma_{ij}/r)^{12} - (\sigma_{ij}/r)^6)$, with the same depth ϵ for tracer -solvent and solvent-solvent interactions. Figure 3 presents the results obtained for size ratios σ_1/σ_2 between 1 and 20. As σ_1/σ_2 increases, the tracer diffusion constant first increases, then saturates and for σ_1/σ_2 larger than about 10 becomes independent of tracer size. Small particles can therefore be considered as pointlike objects moving in a frozen energy landscape created by the solvent particles.

This is confirmed by a study of the "Einstein frequency distribution" for the small particles, defined in the following way: at every time step, the three eigenvalues of the matrix of second derivatives of the potential

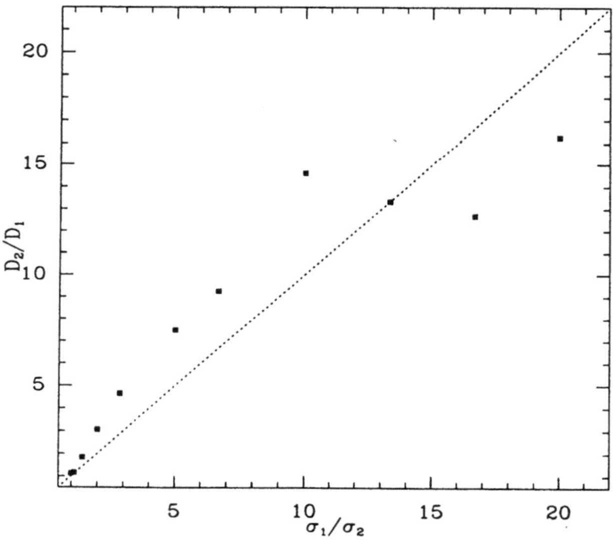

Ratio of the tracer diffusion constant to the self diffusion constant of the solvent as a function of the size ratio σ_1/σ_2. The Stokes-Einstein prediction, $D_2/D_1 = \sigma_1/\sigma_2$, is indicated by a the straight line.

felt by the tracer particle are computed. Dividing these eigenvalues by the particle mass, one obtain three "square frequencies" whose average value is the usual Einstein frequency. The distribution of these frequencies, $p(\omega^2)$, provides information on the energy landscape explored by the tracer particle. When the tracer has the same size as the solvent, almost all the eigenvalues are positive. This indicates that the particle remains in a potential well formed by its neighbours, and moves by "pushing" these neighbours. On the other hand, a small particle spends about 30% of its time in regions of negative curvature of the potential, and thus hopping over energy barriers. Most likely, the latter type of motion can be described as resulting of "binary collisions" in the language of kinetic theory [15], while the description of the former requires the inclusion of "mode-coupling" type terms. A detailed study of this effects, as well as the influence of other parameters on tracer diffusion will be presented elsewhere [16]

Acknowledgments

The work described above resulted from collaborations with J-N Roux, J-P Hansen, M.L. Klein, G.F. Signorini and F. Ould-Kadour. The financial support of CNRS is acknowledged.

REFERENCES

1. J-N. Roux, ,J-L. Barrat and J-P. Hansen, J. Phys. Condensed Matter **1**, 7171 (1989)

2. J-L. Barrat, J-N. Roux and J-P. Hansen, to be published in Chemical Physics (December 1990)

3. G. Pastore, B. Bernu,J-P. Hansen and Y. Hiwatari, Phys. Rev. **38**, 454 (1988)

4. B. Bernu, J-P. Hansen, Y. Hiwatari and G. Pastore, Phys. Rev. **A36**, 4891 (1987)

5. H. Miyagawa, Y. Hiwatari, B. Bernu and J-P. Hansen, J. Chem. Phys. **88**, 3879 (1988)

6. M. Goldstein, J. Chem. Phys. **51**, 3278 (1969)

7. E. Rössler, Phys. Rev. Lett. **65**, 1595 (1990)

8. U. Bengtzelius, W. Götze W. and A. Sjölander, J. Phys. **C17**, 5915 (1984).

9. W. Götze and L. Sjögren in *Dynamics of disordered materials*, Richter D., Dianoux A.J., Petry W., Teixeira J. eds. Springer, Berlin (1989)

10. S.P. Das and G.F. Mazenko, Phys. Rev. **A34**, 2265 (1986)

11. G.F. Signorini, J-L. Barrat and M.L. Klein, J. Chem. Phys. **92**,1294 (1990)

12. W. Knaak, F. Mezei and B. Farago, Europhys. Lett. **7**, 529 (1988)

13. F.S. Howell, R.A. Bose, P.B. Macedo, and C.T. Moynihan, J. Phys. Chem. **78**, 639 (1974)

14. G.L. Pollack and J.J. Enyeart, Phys. Rev. ,**31**, 980 (1985)

15. J-P. Hansen and I.R. McDonald, *Theory of simple liquids* Academic Press, New-York. (1986)

16. F. Ould-Kadour and J-L. Barrat, in preparation.

Molecular Dynamics Study of Self-Diffusion and the Liquid–Glass Transition

Y. Hiwatari

Department of Physics, Faculty of Science, Kanazawa University, Kanazawa 920, Japan

Abstract. Dynamical properties of supercooled liquids have been studied with simple models for both a binary-alloy fluid and a molten LiI via molecular dynamics simulations. To search for the nature of dynamical slowing down associated with the liquid-glass transition, at the microscopic level, this work focuses on the study of the mechanism of the diffusivity as well as the significant behavior of the atomic motion in highly supercooled liquids close to the transition. Two model fluids have been studied in this work, the binary soft-sphere mixture and LiI, which exhibit common dynamical singular behavior on approaching the respective liquid-glass transition. However, the remarkable difference of the mass ratio of the species between the binary soft sphere mixture and LiI results in different dynamical details from one to the other. For the molten LiI, jump motions of Li ions are well established in both normal and supercooled fluid phases, while for the binary soft-sphere model clear jump motions are only possible in the highly supercooled phase close to the glass transition. It is shown that the residence-time distribution obtained through the molecular dynamics calculation confirms the power-law form of the jump-rate distribution which has recently been proposed by Odagaki and Hiwatari with the trapping diffusion model (Phys. Rev. A 41 (1990) 929).

1. Introduction

The glass transition is not a true thermodynamic phase transition, but a dynamical change associated with a dramatic increase of the relaxation time of density fluctuations as well as viscosity, around which the relaxation time becomes longer than observation times and viscosity is the order of 10^{13} Poise. Thus, it is reasonably accepted that a system quenched into the vicinity of the transition falls out of complete thermodynamic equilibrium, at least in time scales under observation. We note that the dynamical properties, driven by such slow dynamics, are essential and the key to understanding the nature of the liquid-glass transition. Therefore, a molecular dynamics (MD) study is very useful for these considerations. Since we are more concerned with fundamental and general problems about characterization of the dynamical singularities which take place near the glass transition rather than properties of a particular glass material, here we consider simple model fluids for which MD simulations were carried out.

It is well known that a binary fluid is more easily transformed into a glassy state than a monatomic one-component fluid /1,2/. Simple fluids such as argon-like liquids have not been transformed into a glassy state by laboratory experiments so far. The existence of the glass transition for such simple fluids has only been examined by computer experiments, either molecular dynamics or Monte Carlo simulations, in which the rate (velocity) of quenching is roughly 4 to 6 orders of magnitude faster than typical experimental quench rates. However, the quenched states obtained by computer experiments for simple fluid models are not very stable; they are often immersed in spontaneous nucleations during long time steps of simulations. In order to overcome such a problem and study the dynamical properties of highly supercooled states, in particular to study slow dynamics characteristic to the liquid-glass transition in which we are particularly interested, we should better consider binary fluid models, in which two different species are mixed.

Dynamical properties may significantly depend on the mass ratio of the heavier to the lighter species of a given two-species fluid. For the binary soft-sphere mixture, modeled for real binary alloys, reasonable values of the mass ratio range between one and two, while in a molten LiI the mass ratio of iodine to lithium ions yields about 18.3, which is much larger than the former case /3/. Such a remarkable difference of the mass between iodine and lithium ions, combined with the Coulomb attraction between unlike ions, may give us an idea for the diffusion mechanism of the lithium ions to follow activated processes, which, as we will see below, turns out to be quite reasonable for both normal and supercooled molten LiI, while such a model becomes unreasonable for normal binary soft sphere fluids, being only reasonable in highly supercooled fluids close to the glass transition.

It has been shown by the molecular dynamics simulations for the binary soft-sphere model that on approaching the liquid-glass transition, cooperative (concerted) jump motions are significant, which means that several near-neighbor atoms jump at nearly the same time and permute their positions /4/. These jump motions are essential to the mechanism of the slow dynamics (e.g., the diffusivity) characteristic in the vicinity of the transition. Both mode-coupling and hydrodynamic models for the glass transition, however, do not take into consideration these important local dynamical fluctuations, and consequently are not reliable in the transition region /5,6/. Recently, Odagaki and Hiwatari have proposed a new approach to the problem of the dynamical nature of the glass transition using the trapping diffusion model, based on a jump diffusion model with a stochastic equation for a tracer atom and with the power-law form of the jump-rate distribution function /7,8/. It has been shown that the predictions of the trapping diffusion model, such as non-t(time)-linear (subdiffusive) behavior of the mean square displacement, anomalous diffusion, and stretched exponential decay of the density autocorrelation function, are all in good agreement with the results of the molecular dynamics simulations for the binary soft-sphere model. Under these cimcumstances, it is of particular interest to investigate further the validity of the jump diffusion model from microscopic points of view. For this purpose, in this work, by analysing the data obtained by the molecular dynamics simulations we have computed the jump-rate or equivalently residence-time distribution to see to what extent the power-law jump-rate distribution is reliable, which was assumed by Odagaki and Hiwatari in their trapping diffusion model for the glass transition. We will also study the common nature of the anomalous dynamical properties near the glass transition for both the binary alloy and LiI, as well as their differences.

2. The Models

We first consider the binary soft-sphere model, for which the interatomic potentials are given as follows:

$$v_{\alpha\beta}(r) = \epsilon(\sigma_{\alpha\beta}/r)^{n}, \tag{1}$$

$$\sigma_{\alpha\beta} = (\sigma_{\alpha}+\sigma_{\beta})/2, \tag{2}$$

where α and β denote species indices (=1, 2), the exponent n was taken to be 12 throughout this work, and the additivity rule was assumed for the interaction between different species. The core-size ratio σ_2/σ_1 was taken to be 1.2, and the mass ratio m_2/m_1 to be 2. The concentration of the species 1, x_1 ($x_1=N_1/N$ and $x_2=N_2/N$ with $N=N_1+N_2$) was varied between 0.25 and 1.0. Here we present only the equimolar case, namely $x_1=0.5$, which was most extensively studied. According to the scaling properties of the soft-sphere model /1,2/, we have only the single coupling constant Γ, defined below in Eq. (3), provided that x_1 and σ_2/σ_1 are fixed. It is our intention to use the effective one-component approximation (conformal solution theory) so that the variation of both x_1 and σ_2/σ_1 can be effectively coupled into Γ. Consequently we have a single effective coupling constant Γ_{eff}:

$$\Gamma = (N\sigma_1^{\ 3}/V)(\varepsilon/k_BT)^{1/4} , \tag{3}$$

$$\sigma_{eff}^{\ 3} = \sum_{\alpha}\sum_{\beta} x_\alpha x_\beta \sigma_{\alpha\beta}^{\ 3} , \tag{4}$$

$$\Gamma_{eff} = \Gamma(\sigma_{eff}/\sigma_1)^3 . \tag{5}$$

This approximation makes it possible to define any thermodynamical equilibrium state in terms of the single effective coupling constant Γ_{eff}. For simplicity we use Γ_{eff} instead of Γ throughout this paper. In fact, it can been shown that the equation of states for the binary soft-sphere fluid over a wide range of concentrations and for different core size ratios can be well scaled on a single curve with Γ_{eff}. The freezing point is given by $\Gamma_{eff}=1.15$. By the series of the molecular dynamics studies for both one- and two-component soft-sphere fluids, it has been shown that the liquid-glass transition occurs around $\Gamma_{eff}=1.58$. The simulation was carried out on a periodic box containing N=500 atoms with the standard Gear algorithm (7th order force representation) and Nosé's constant temperature techniques /9/.

In molecular dynamics simulations for molten LiI, we have used the following pair potentials /10/:

$$v_{\alpha\beta}(r)=z_\alpha z_\beta e^2/4\pi\varepsilon_0 r + (1+z_\alpha/n_\alpha+z_\beta/n_\beta) \tag{6}$$
$$\times\ b\ \exp\{(\sigma_\alpha+\sigma_\beta-r)/\rho\} - c_{\alpha\beta}/r^6 - d_{\alpha\beta}/r^8$$

where the values of the parameters were taken from ref. 11 with a slight modification for the parameter ρ, reduced by a factor 0.92 of the original value. The molecular dynamics simulations were carried out with, in the largest case, 1024 respective ions using Ewald calculations of Coulomb forces between ions (with either the constant-energy or constant-temperature method). Details of the simulations will be published in our forthcoming paper.

3. Results and Discussion

Figure 1 shows the power spectra of the velocity autocorrelation function for supercooled molten LiI at

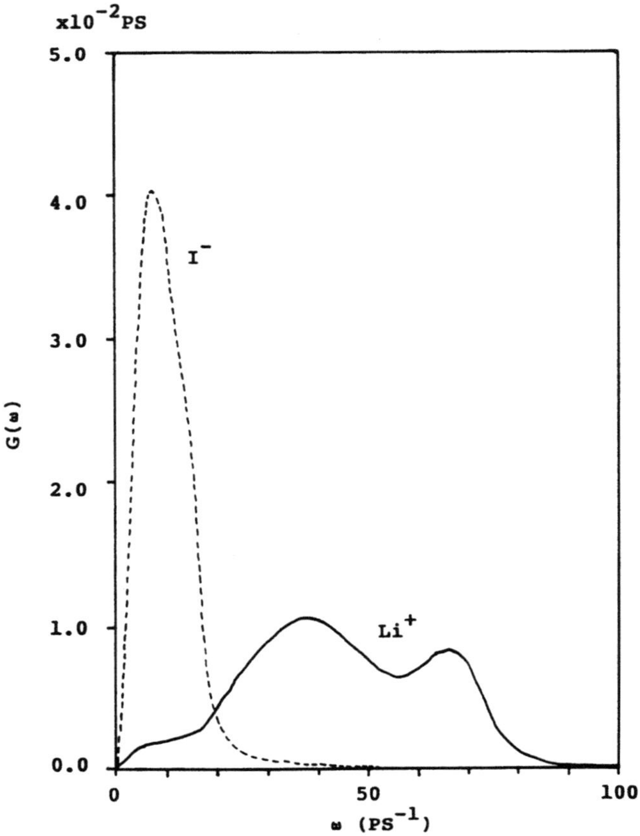

Fig. 1. Power spectra of lithium and iodine ions in supercooled molten LiI at 400K.

T=400K. There is a significant difference in behavior between the curves for lithium and iodine ions: The spectrum for the lithium ion is much more widely spread over high frequencies than that for the iodine ion. The two peaks in the spectrum of the lithium ion, the low-frequency and the high-frequency peaks, correspond to the transverse and longitudinal optical phonon modes, respectively (see Fig. 2). A much stronger single peak for iodine is rather an apparent one, which results from superposition of damped acoustic low-freqency phonon modes and local vibration modes of atoms or ions; the peak frequency is generally smaller than the correponding Einstein frequency. A similar result was obtained for the binary soft-sphere mixture, in which the power spectrum for different species has a rather similar form /2/. Thus, the difference in the behavior of the power spectrum between both fluids is remarkable.

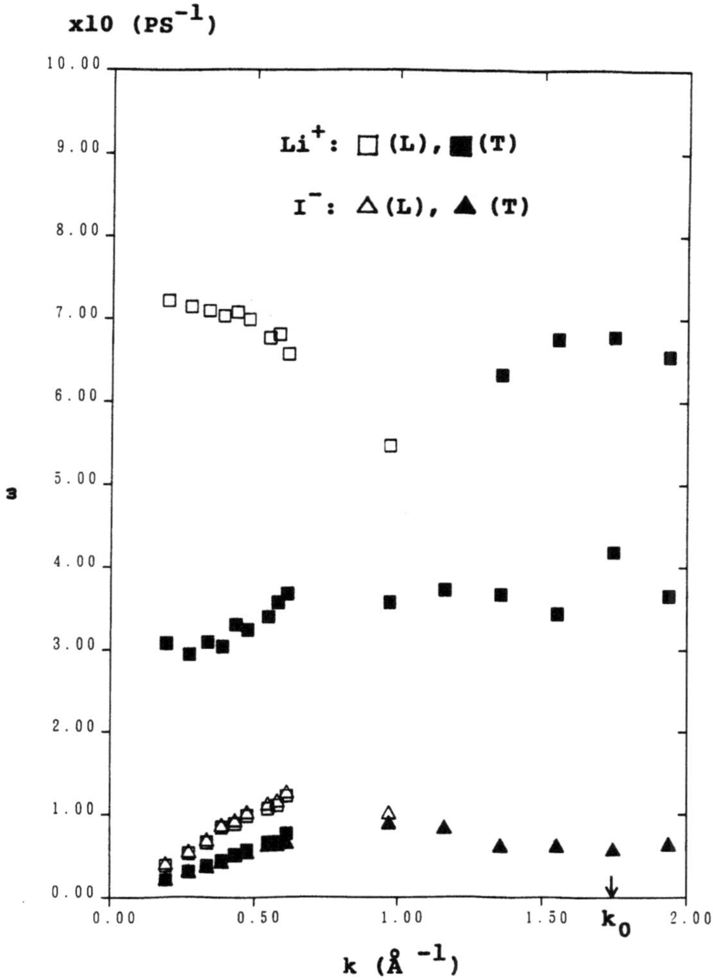

Fig. 2. Dispersion relations ω vs k for supercooled molten LiI at 400K, as calculated from the maximum location of the Fourier spectrum of the longitudinal (L) and transverse (T) partial (lithium- and iodine-) current correlation functions. k_0 denotes the maximum location of the static total structure factor at the same temperature.

Mean square displacement (MSD) $R_2(t)$ for highly supercooled states hardly reaches a t(time)-linear behavior as expected from the theory of Brownian motion, at least on the time scale of the molecular dynamics simulations (see Fig. 3), exhibiting a remarkable contrast to fluid states far from the glass transition. The apparent diffusion constant, as

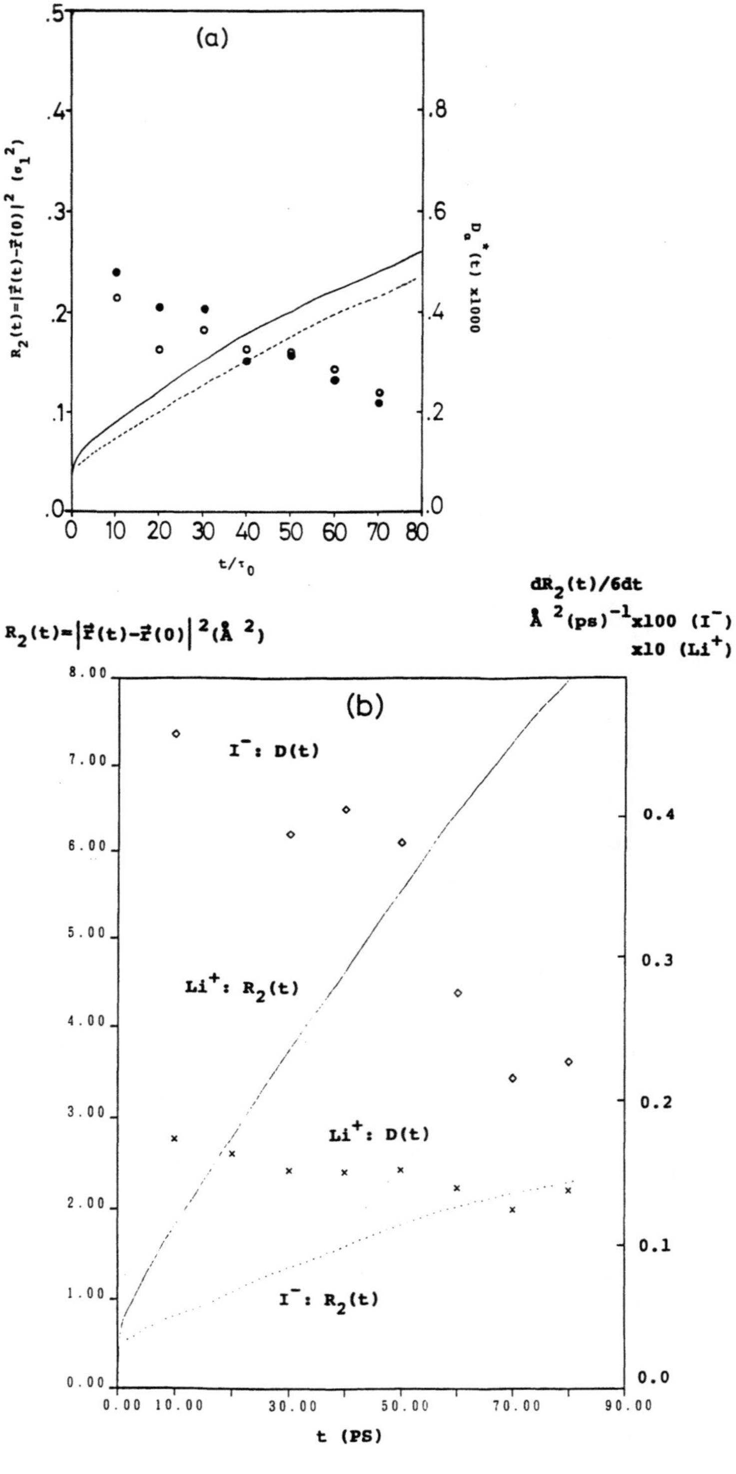
(a)
$R_2(t)=|\vec{r}(t)-\vec{r}(0)|^2\ (\sigma_1^2)$
$D_a^*(t)\ x1000$
t/τ0
(b)
$R_2(t)=|\vec{r}(t)-\vec{r}(0)|^2$(Å 2)
$dR_2(t)/6dt$
Å 2(ps)-1x100 (I-)
x10 (Li+)
I-: D(t)
Li+: R2(t)
Li+: D(t)
I-: R2(t)
t (PS)

Fig. 3. Time variation of the mean square displacement $R_2(t)$ and temporal diffusivity $D_\alpha(t)=dR_2(t)/6dt$. The results for the binary soft-sphere model at $\Gamma_{eff}=1.55$ are shown in (a), where the solid and dotted curves represent $R_2(t)$ for the lighter and heavier species, respectively, while solid and white circles represent the reduced diffusion constant $D^*_\alpha(t)$ for the lighter and heavier species, respectively. $\tau_0=\tau(T^*)^{-1/2}(\rho^*)^{-1/3}$, $\tau=\sqrt{m_1\sigma_1{}^2/\varepsilon}$, and $D^*_\alpha(t)=D_\alpha(t)\tau(T^*)^{-5/12}$ with $T^*=k_BT/\varepsilon$ and $\rho^*=(N\sigma_1{}^3/V)$. In (b) are shown those for supercooled molten LiI at 400K. The symbols **x** and ◇ denote the temporal diffusivity as defined above for lithium and iodine ions, respectively.

calculated from the derivative of the MSD with respective to time t, $D_\alpha(t)=dR_2(t)/6dt$, tends to decrease with increasing time, thus an accurate diffusion constant is hardly obtained. It is found that the time variation of the MSD can be reasonably well fitted by the power-law form $R_2(t)\propto t^\beta$ $(0<\beta<1)$. The exponent β is equal to one in the normal to moderately supercooled fluid range, while it becomes considerably less than one at lower temperatures (see Fig. 4). It has also been seen that the incoherent scattering function (the self-part of the density autocorrelation function) shows not a simple exponential decay, but a stretched exponential decay /12,13/, which implies the significance of non-Gaussian dynamics for highly supercooled fluid states as well as the glassy states.

In order to measure non-Gaussian behavior of the system, we have computed the so-called non-Gaussian parameter (NGP), which is defined as

$$A_\alpha(t) = \frac{3}{5}\frac{\langle|\vec{r}(t)-\vec{r}(0)|^4\rangle}{\langle|\vec{r}(t)-\vec{r}(0)|^2\rangle^2} - 1 \qquad (7)$$

It is known that the NGP exactly vanishes in some ideal systems such as gases, harmonic solids, Brownian systems, and so on. In the normal fluid phase, the NGP yields a non-vanishing but small value in the initial time domain of the order of the Einstein period, but it relaxes very quickly to zero for longer times. However, in the highly supercooled fluid phase, the NGP deviates remarkably from zero, exhibiting a clear maximum at a certain time; the decay of $A_\alpha(t)$ becomes more and more slow and the time at which a maximum appears becomes longer and longer as the glass transition is approached (see Fig. 5). This result implies that mechanisms of the diffusion in such highly supercooled fluid states significantly differ from usual Brownian motions, that

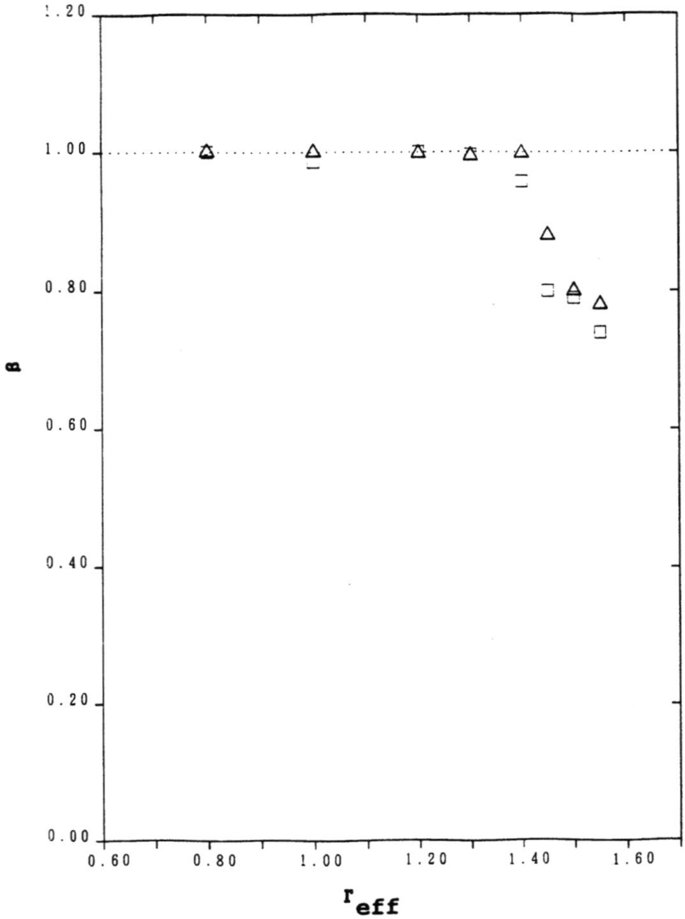

Fig. 4. β vs Γ_{eff}, as calculated by power-law fits for the mean square displacement, i.e., $R_2(t) \propto t^{\beta}$, (binary soft-sphere model). Triangles are for the lighter species, while squares are for the heavier species.

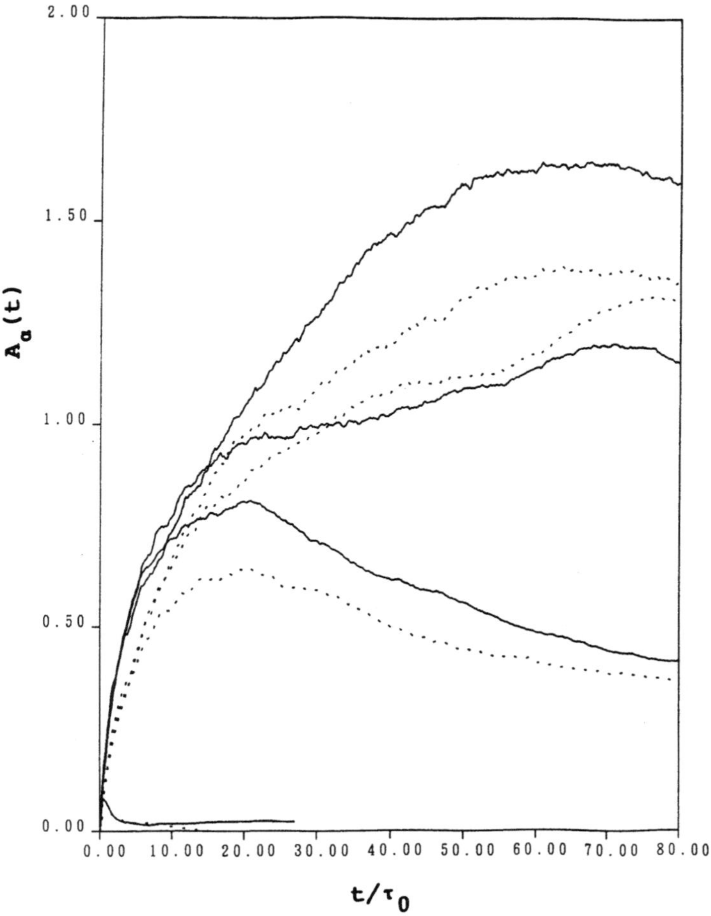

Fig. 5. Non-Gaussian parameters $A_\alpha(t)$ for the binary soft-sphere fluids. Solid curves are for the lighter species, while dotted curves are for the heavier species at the corresponding Γ_{eff} : Γ_{eff}'s are 0.8, 1.45, 1.50, and 1.55, respectively, from the bottom solid curve to the top solid curve.

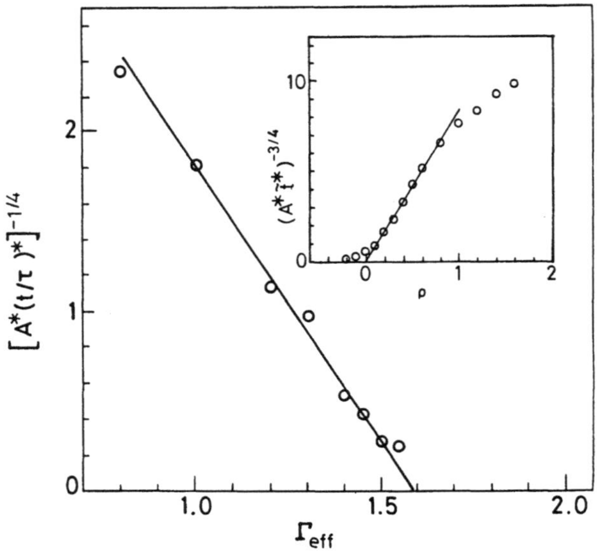

Fig. 6. The product of the maximum A^* of the non-Gaussian parameter and the corresponding time t^* for the binary soft-sphere model behaves as if it diverges at $\Gamma_{eff}=1.58$, indicating that the glass transition (Gaussian-to-non-Gaussian transition) takes place there. We observe that $A^*(t/\tau)^* \propto (\Gamma_g - \Gamma_{eff})^{-4}$. The inset is a similar plot for the stochastic trapping model, where the transition takes place at $\rho=0$ and $A^*\tilde{t}^*$ behaves as $\rho^{-4/3}$ near the transition point.

is, dynamical correlations between successive jumps are not negligible. Direct observations of correlated (concerted) jump motions have been reported by Miyagawa et al. via molecular dynamics calculations (see Introduction and ref. 4). We also noted that molten LiI shows stronger correlations, which is presumably due to long range Coulombic interactions between the ions.

The temperature dependence of the product of the time and the value at the maximum location of the the NGP shows a critical behavior as the temperature is lowered, with which one can determine the glass transition temperature T_g (see Fig. 6), as proposed by the trapping diffusion model /7,8/. This method benefits an accurate determination of T_g, since the maximum of the NGP occurs in an intermediate time range which is accessible by the molecular dynamics simulation. Figure 6 shows a comparison between the molecular dynamics result for the binary soft-sphere model and the trapping diffusion model. In the latter model, the glass transition occurs at $\rho=0$, where the diffusivity vanishes. From this result, we obtain that

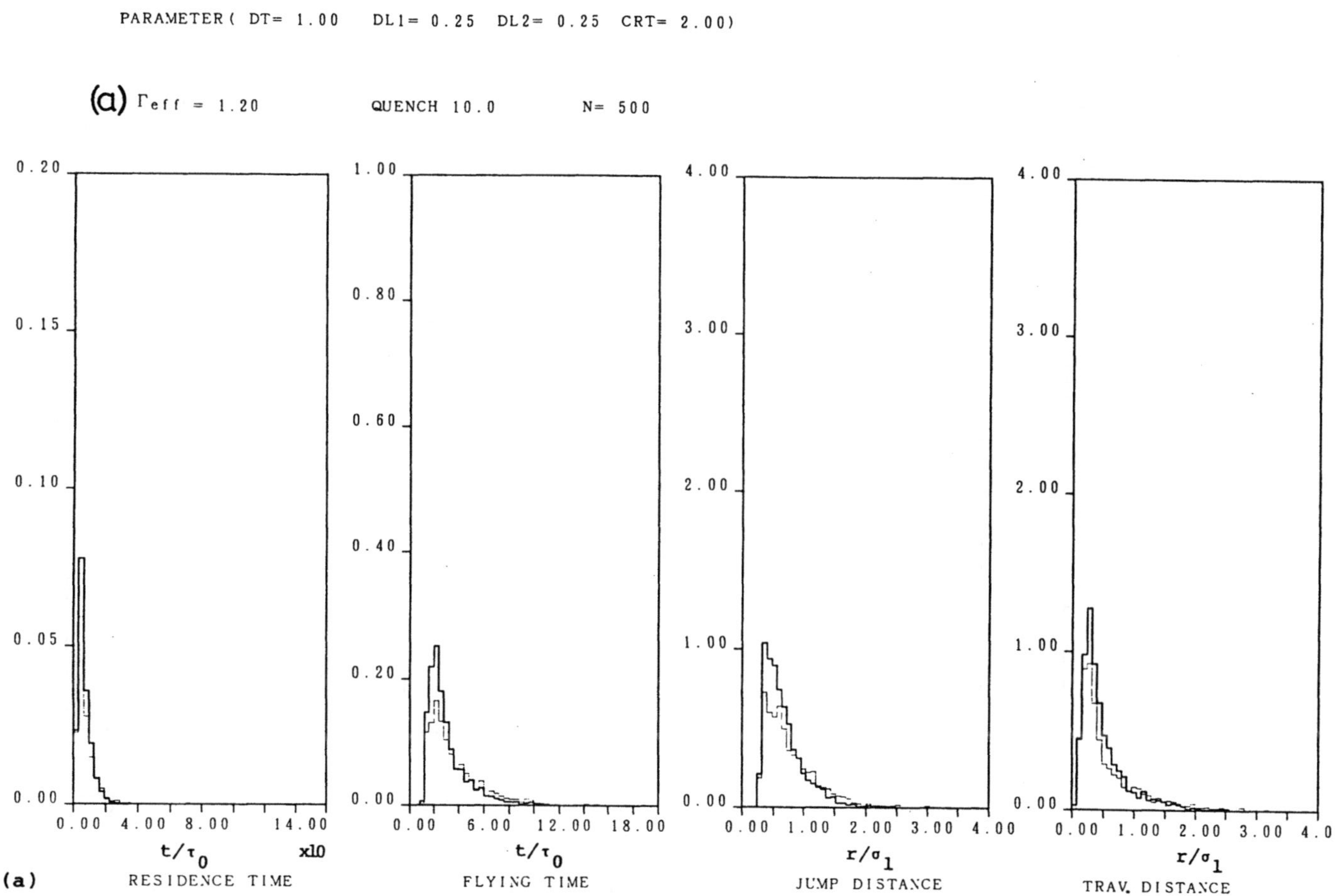

PARAMETER (DT= 1.00 DL1= 0.25 DL2= 0.25 CRT= 2.00)
(a) Γeff = 1.20
QUENCH 10.0
N= 500
0.20
0.15
0.10
0.05
0.00
0.00 4.00 8.00 14.00
t/τ0
x10
RESIDENCE TIME
1.00
0.80
0.60
0.40
0.20
0.00
0.00 6.00 12.00 18.00
t/τ0
FLYING TIME
4.00
3.00
2.00
1.00
0.00
0.00 1.00 2.00 3.00 4.00
r/σ1
JUMP DISTANCE
4.00
3.00
2.00
1.00
0.00
0.00 1.00 2.00 3.00 4.00
r/σ1
TRAV. DISTANCE

Fig. 7(a)

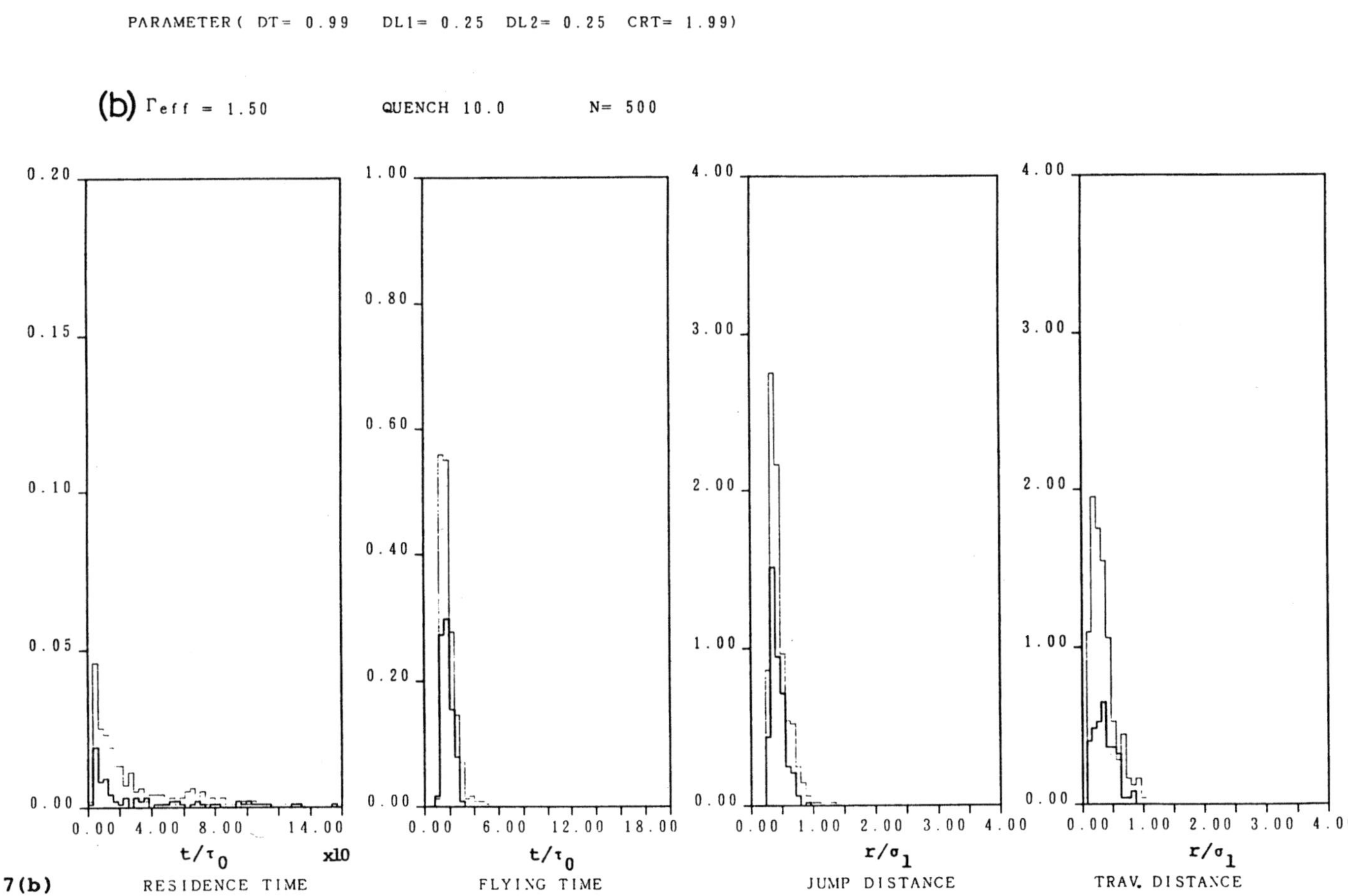
PARAMETER (DT= 0.99 DL1= 0.25 DL2= 0.25 CRT= 1.99)
(b) Γ_{eff} = 1.50
QUENCH 10.0
N= 500
0.20
0.15
0.10
0.05
0.00
0.00 4.00 8.00 14.00
t/τ_0
x10
RESIDENCE TIME
1.00
0.80
0.60
0.40
0.20
0.00
0.00 6.00 12.00 18.00
t/τ_0
FLYING TIME
4.00
3.00
2.00
1.00
0.00
0.00 1.00 2.00 3.00 4.00
r/σ_1
JUMP DISTANCE
4.00
3.00
2.00
1.00
0.00
0.00 1.00 2.00 3.00 4.00
r/σ_1
TRAV. DISTANCE

Fig. 7(b)

$$\rho \propto (\Gamma_g - \Gamma_{eff})^3 \qquad (8)$$

and the glass transition for the binary soft-sphere model is estimated to be Γ_g =1.58. We note that Γ_g is substantially larger than the value Γ_{eff} =1.45, beyond which non-t-linear behavior of the MSD and stretched exponential decay of the incoherent scattering function are observed (see Fig. 4). It is reasonably acceptable that the latter behaviors are precursors of the liquid-glass transition and these properties may substantially depend on the time scales under consideration, as we have mentioned above in the apparent diffusivity, which tends to decrease with increasing time.

Details of the jump motions of atoms in supercooled fluid states were obtained by monitoring the trajectory of each atom with the molecular dynamics data. We have computed the jump-distance (JD), flying-time (FT) and residence-time (RT) distributions. The results are shown in Fig. 7. A sharp peak is observed in the distributions of both the JD and FT for the highly supercooled binary soft-sphere fluid (Γ_{eff} =1.50), while the corresponding distributions become much broader for Γ_{eff} =1.2 (close to freezing). This means that the diffusion processes in the former can be reasonably well described by a single-distance jump model. Similar results were obtained for molten LiI. In this case, however, each distribution exhibits a comparably sharp peak even for T=760k (near the freezing temperature) as well as for T=400K (quenched state). Such differences in the behavior between both models are remarkable, which originates from the difference of the mass ratios and their potentials.

The residence-time distribution for the binary soft-sphere supercooled state for Γ_{eff} =1.55 is shown in Fig. 8. Here is shown a comparison with the power-law jump-rate distribution:

$$P(w_s) = \frac{\rho+1}{w_0^{\,\rho+1}} w_s^{\rho}: \qquad 0<w_s<w_0, \quad -1<\rho. \qquad (9)$$

Fig. 7. The residence-time(RT), flying-time(FT), jump-distance(JD) and traveling-distance (the distance displaced during a residence time) histograms for the binary soft-sphere model for (a) Γ_{eff} =1.20 and (b) Γ_{eff} =1.50. Solid histograms are for the lighter species, while light-solid ones are for the heavier species.

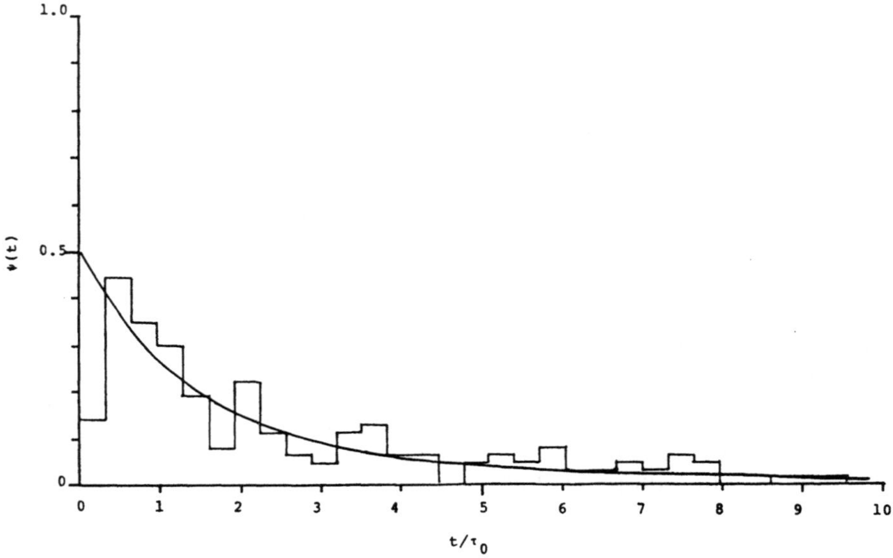

Fig. 8. Comparison of the residence time distribution between the binary soft-sphere model for $\Gamma_{eff}=1.55$ (histogram) and the trapping model for $\rho=0.01$ with $zw_0\tau_0=0.099$ (solid line).

The probability that an atom at site s makes the first jump between t and t+dt is given by

$$\psi(\vec{s},t)dt=zw_s e^{-zw_s t}\, dt. \qquad (10)$$

Therefore, the residence (waiting) time distribution becomes

$$\psi(t)=\int_0^{w_0} \frac{\rho+1}{w_0^{\rho+1}}\, w_s^{\rho}\, zw_s\, e^{-zw_s t}\, dw_s. \qquad (11)$$

Putting $x=zw_0t$, we obtain

$$\psi(t)=\frac{(\rho+1)zw_0}{(zw_0t)^{\rho+2}}\int_0^{zw_0t} x^{\rho+1}\, e^{-x}\, dx. \qquad (12)$$

Using a scaled time $\tilde{t}=zw_0t$ and the Tricomi's incomplete gamma function

$$\gamma^*(a,x)= \frac{x^{-a}}{\Gamma(a)} \int_0^x t^{a-1} e^{-t}\, dt, \qquad (13)$$

we can rewrite Eq. (12) as

$$\psi(\tilde{t})d\tilde{t} = (\rho+1)\Gamma(\rho+2)\ \gamma^*(\rho+2,\tilde{t})d\tilde{t}. \qquad (14)$$

In order to compare this with the residence-time distribution obtained by the molecular dynamics calculation, the parameters ρ and zw_0 were determined by the method of least square fits. For $\Gamma_{eff}=1.55$ in Fig. 8, the result of the best fittings yielded $\rho=0.01$ and $zw_0\tau_0=0.099$, and we have obtained a satisfactory agreement between the two. Thus, we conclude that the assumption of the power-law form for the jump-rate distribution has been found to be reasonable at least for highly supercooled fluid states in simple models.

The trapping diffusion model predicts a sharp liquid-glass transition in the infinite time scale similar to an ideal glass transition proposed by mode-coupling models. However, the trapping diffusion model also predicts that, on intermediate time scales compatible with those of the molecular dynamics simulations, the sharp transition disappears and becomes a smeared one, which shows good agreements of various dynamical singular behaviors in the vicinity of the glass transition with the molecular dynamics results for the binary soft-sphere model.

Acknowledgements

This work was partly supported by a Grant-in-Aid from the Ministry of Education, Science and Culture. Computations were carried out with the supercomputers at Kyoto University, the Plasma Institute of Nagoya University, and the Institute of Molecular Science. The author appreciates CPU allocation on these supercomputers. He also thanks T. Odagaki and H. Miyagawa for valuable comments and discussions.

References

1) B. Bernu, Y. Hiwatari, and J. P. Hansen, J. Phys. C **18**, L371 (1985).
2) B. Bernu, J. P. Hansen, Y. Hiwatari, and G. Pastore, Phys. Rev. **A 36**, 4891 (1987).
3) Y. Hiwatari, H. Miyagawa, and S. Itoh, Proc. of Int. Workshop on Computational Materials Science, August 23-24, 1990, Tsukuba Science City, Ibaraki, Japan, p. 57.

4) H. Miyagawa, Y. Hiwatari, B. Bernu, and J. P. Hansen, J. Chem. Phys. **88**, 3879 (1988); H. Miyagawa and Y. Hiwatari, Phys. Rev. A 40, 6007 (1989).
5) E. Leutheusser, Phys. Rev. **A 29**, 2765 (1984).
6) S. P. Das and G. F. Mazenko, Phys. Rev. **A 34**, 2265 (1986); Phys. Rev. **A 36**, 211 (1987).
7) T. Odagaki and Y. Hiwatari, Phys. Rev. **A 41**, 929 (1990); J. Non-Cryst. Solids 117/118, 887 (1990).
8) T. Odagaki and Y. Hiwatari, Phys. Rev. **A 43**, 1103 (1991).
9) S. Nosé, J. Chem. Phys. **81**, 511 (1984).
10) S. Itoh, M. Konagai and K. Takahashi, Z. Naturforsch. **46a**, 155 (1991).
11) M. P. Tosi and F. G. Fumi, J. Phys. Chem. Solids, **25**, 31 (1964); ibid., **25**, 45 (1964).
12) G. Pastore, B. Bernu, J. P. Hansen, and Y. Hiwatari, Phys. Rev. **A 38**, 454 (1988).
13) Y. Hiwatari and H. Miyagawa, J. Non-Cryst. Solids **117/118**, 862 (1990).

Molecular Dynamics Study of Fluctuation and Relaxation in Disordered Systems – Liquid and Glass

F. Yonezawa[1] *and S. Sakamoto*[2]

[1]Department of Physics, Faculty of Science and Technology, Keio University, 3-14-1 Hiyoshi, Kohokuku, Yokohama 223, Japan
[2]Department of Applied Physics, Seikei University, 3-3-1 Kichijoji-Kitamachi, Musashinoshi, Tokyo 180, Japan

Abstract. By means of the constant–pressure molecular dynamics simulations, we construct various disordered states of a Lennard–Jones system in order to study the characteristic features of different disordered structures. In particular, we study the atomic configurations of a supercooled liquid and a glassy solid by investigating the details of fluctuation and relaxation in these systems. To this end, we introduce several microscopic structure parameters which clarify some essential aspects of disordered systems.

1. Introduction — Purpose, Model and Method

The purpose of the present article is to achieve a better understanding of the essential nature of amorphous structures by studying the ways in which various kinds of disordered systems in the metastable states are relaxed to the stable states across activation barriers [1].

As examples of such disordered systems, we construct a supercooled Lennard–Jones liquid and a Lennard–Jones glass by molecular dynamics (MD) simulations. To be more precise, we choose a model system composed of 864 particles interacting via the 12–6 Lennard–Jones potentials, and place these atoms in a simulation cell with periodic boundary condition.

For this model system, we carry out simulations [1,2,3] by making use of the constant–pressure MD method first demonstrated by Damgaard Kristensen [4] and later neatly formulated by Andersen [5]. The outcomes from a computer simulation can be closely compared with the experimental results when the constant–pressure technique is adapted, because most experiments about liquids and glasses are performed under a constant (atmospheric) pressure.

In order to make the comparisons easier, we use, in our simulations, dimensional variables by scaling energy in ε, length in σ, mass in M, temperature in ε/k_B, pressure in ε/σ^3, and time in $(M\sigma^2/\varepsilon)^{1/2}$, where ε and σ are the potential parameters of the Lennard–Jones potential, M is the atomic mass, and k_B is the Boltzmann constant. It is widely accepted that the Lennard–Jones potential applies very well to rare gases. The magnitudes of temperature ε/k_B are respectively 118 K, 170 K and 244 K for argon, krypton and xenon.

In what follows, we make detailed analyses of a supercooled liquid at T=0.46 and a glass at T=0.34 which are produced from our simulations. The number N of particles in each system is 864.

In order to study the behaviours of the these two metastable states, we relax these systems by keeping the temperatures of the systems at their respective original values (at T=0.46 and T=0.34). We refer to this process as *'annealing.'* Under the process of annealing, we observe the changes of several physical quantities of these systems as functions of the annealing time.

The situation is explained in Fig. 1 for the volume at T=0.34. The phase diagrams are generally drawn on a plane of a given variable versus temperature such as on the V–T plane, while, in our case, we add the

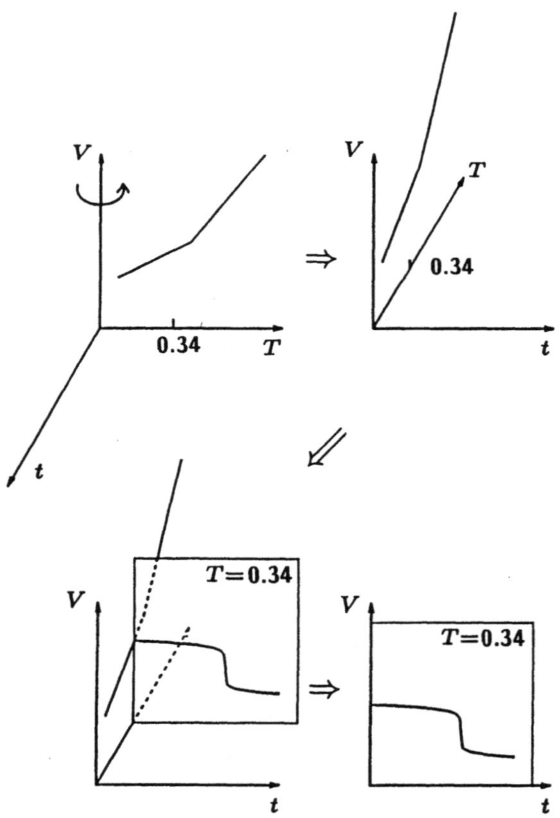

Fig. 1. Three axes, time t, temperature T, and volume V, which show the concept of annealing.

time axis as the third axis. Then it becomes possible to define a plane of a given variable versus time such as the $V-t$ plane at each temperature (for example, at T=0.34). The effects of annealing is discussed on this new plane.

2. PDF through the Annealing Process

The pair distribution function (PDF) $g(r)$ is a most convenient quantity for examining the atomic configurations. Time step is taken to be 5×10^{-15} second. Figure 2 demonstrates the change of the PDF at various time steps through the annealing process of a supercooled liquid at T=0.46.

At t=1000, the PDF has a second peak of plateau type. This shape of the second peak is very similar to the shape characteristic of glass structures. In fact, it is not only the second peak but also the whole PDF at t=1000 as well as at t=3000 that resemble the corresponding behaviours of glasses, which implies that the atomic configuration in a supercooled liquid has some features typical of a glass.

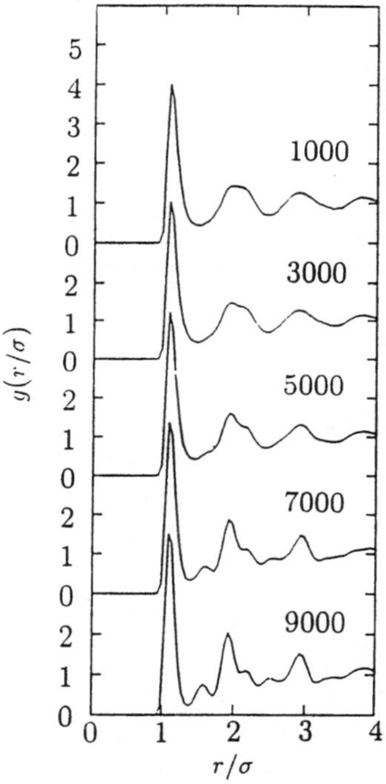

Fig. 2. The PDF $g(r)$ at T=0.46 in the processes of annealing. From top to bottom, t=1000, 3000, 5000, 7000, and 9000.

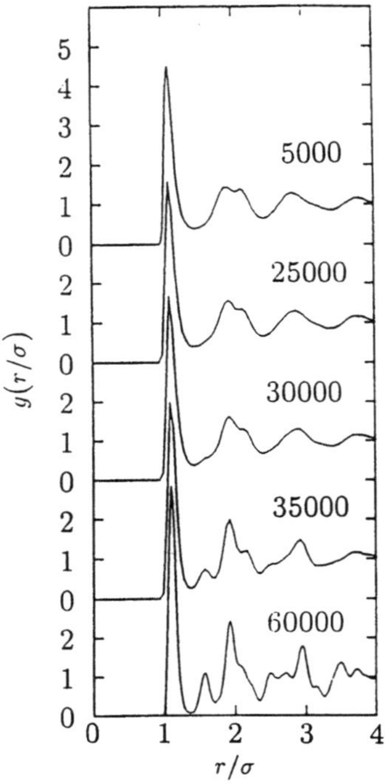

Fig. 3. The PDF $g(r)$ at T=0.34 in the processes of annealing. From top to bottom, t=5000, 25,000, 30,000, 35,000, and 60,000.

As the annealing proceeds, the peak at $\sqrt{3}a$, a being the interatomic distance, becomes apparent. The appearance of this peak is regarded as reflecting the growth of the close–packed structure of a layer type because the distance $\sqrt{3}a$ corresponds to the second–neighbour positions in a close–packed triangular lattice of two dimension.

The peak at $\sqrt{2}a$ becomes discernible at t=5000 time steps, and gradually develops until it is undoubtedly distinct at t=9000 time steps. This distance is characteristic of two atoms that are placed respectively in the succeeding two layers of the close–packed triangular lattices, one layer being stacked quite closely on the other.

The peaks at $2a$ and at $\sqrt{5}a$ also show up as distinct peaks as the atomic structure in the system relaxes on annealing.

All these results prove that, as a consequence of annealing at T=0.46, the atomic distribution achieves a structure with long–range order that is most precisely described by a stacking of two–dimensional close–packed layers.

The time evolution of the PDF on the annealing of a glass at T=0.34 is shown in Fig. 3. Since the starting structure is a glass, it is as expected that the splitting of the second peak is observed at t=5000 time steps, which is the first stage of the annealing. From a close examination of the change in the shapes of the second peak, it turns out that the small subpeaks become recognized at t=30,000 time steps at distances corresponding to the two peaks at $\sqrt{3}a$ and $2a$ characteristic of the crystalline structures, and these subpeaks eventually become more distinct as annealing proceeds.

The intensity of the lower subpeak at $\sqrt{3}a$ increases, indicating the appearance of close–packed configurations of layer type, as is the case for annealing of a supercooled liquid. The intensity of the higher subpeak at $2a$, on the other hand, decreases as t increases while the peak itself grows to be more distinguished. This fact implies that the atoms are distributed more randomly in a glass before the annealing and accordingly here are wider possibilities for the atomic distances around $2a$ when compared with the situation after the annealing.

The time development of the peak at $\sqrt{2}a$ also shows a behaviour similar to that in Fig. 2.

Here again, the mentioned facts serve as the strong evidence for the crystallization of the system on the annealing of a glass at T=0.34.

3. Projections of Atoms

In the preceding section, it has been shown, by the analysis of the time development of $g(r)$, both a supercooled liquid and a glass finally transform into crystals as a result of structure relaxation on annealing. The relaxation of these systems from the metastable states with disordered structures to the stable states with long–range order is observed more directly by drawing the projections of all atoms in these systems onto some appropriate places.

In Fig. 4, are shown the projections of all atoms at every 1000 time steps where the volume of the system is normalized in order to make the comparison easier.

The system is undoubtedly disordered at the first stage of the annealing, as seen from the projection at t=2000. The local nucleations are clearly identified at t=5000, and the ordered (or crystalline) regions coexist with the disordered (or still liquidlike) regions between t=5000 and t=6000.

Between t=6000 and t=8000, a rather large–scale adjustment takes place so that the mismatchings of the symmetry axes among different ordered regions are removed, and, after t=8000, the long–range order extends over the whole system, which becomes all the more evident at t=10,000.

The projections of all atoms in the process of annealing a glass at 40 K are shown in Fig. 5 at every 1000 time steps after t=32,000.

The system already contains regions with some kinds of medium–range order even at t=32,000. From the forms of the projected positions, these

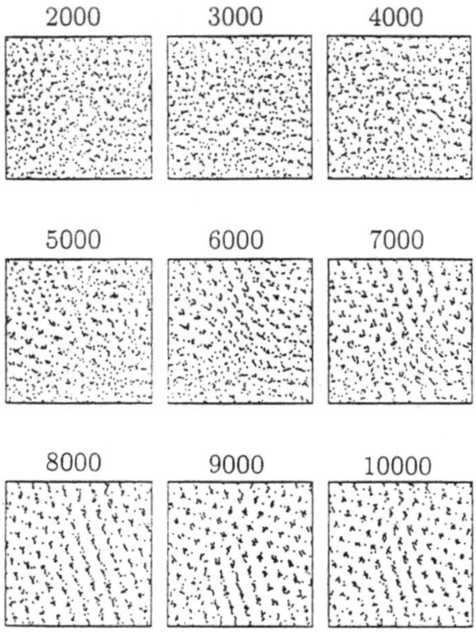

Fig. 4. Projections of atoms onto the xy plane at various time steps at T=0.46. Every 1000 time steps between t=2000 and t=9000.

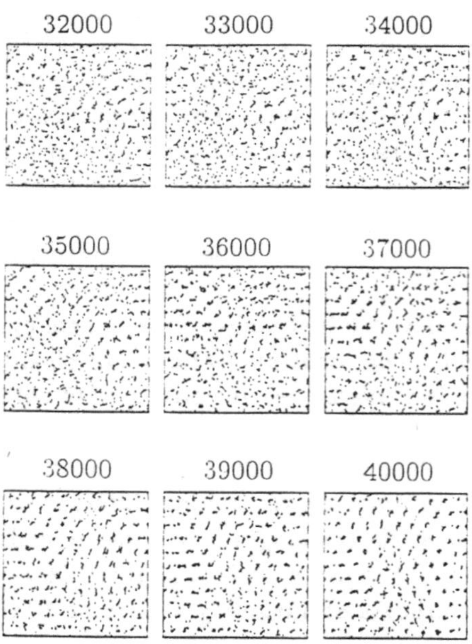

Fig. 5. Projections of atoms onto the xy plane at various time steps at T=0.34. Every 1000 time steps between t=32,000 and t=40,000.

regions are considered to consist of clusters with the structures of crystalline nature. As time proceeds, these regions grow to be crystalline domains, as seen from the projections between t=34,000 and t=37,000. The mismatchings of the crystalline axes among different domains are gradually lifted, but the pace is very low. Even at t=40,000, there still remains a region in which the deviations from the regular lattice points are fairly large.

This tendency probably occurs because the atoms are packed more closely as compared with the situation realized in the annealing process at T=0.46, so a large–scale adjustment of atomic configurations is unlikely to occur. This close packing is also the reason for the fact that, in the annealing at T=0.34, it takes much longer than annealing at T=0.46 before complete long–range order prevails in the whole system.

The features demonstrated by the projections in Figs. 4 and 5 will be analyzed systematically in the next section.

4. Microscopic Structure Parameters

It is widely accepted that the systems in metastable states such as a supercooled liquid and a glass are destined to relax into the most stable states with minimum free energy at the temperature under consideration, which are generally expected to be crystals. The detail of the relaxation in each system is determined by the shape and height of the activation energy that the system has to hurdle.

In order to study the microscopic structures of disordered systems such as liquids and glasses, it is convenient to introduce structure parameters beyond $g(r)$. The tessellation of the whole space according to the atomic configuration [6] has been regarded as one promising method of characterizing amorphous structures. Of several ways for the tessellation, the Voronoi tessellation (or identically tessellation by the Wigner–Seitz cells) has turned out to be most useful. One advantage of the Voronoi tessellation is that it tessellates the whole space uniquely.

We introduce the following microscopic parameters defined on the basis of the Voronoi polyhedra.

1 **Free–volume parameter**
The average free–volume can be estimated by the average volume of all Voronoi polyhedra in the system which is identical to the volume per atom V/N where V is the total volume of the system.

2 **Fluctuation parameter**
The root mean square δv of the volumes of the Voronoi polyhedra serves as a measure for local fluctuation.

3 **Local distortion parameter**
One possible quantity to describe the local distortion in the atomic

configuration is the shape parameter of each Voronoi polyhedron defined by

$$\xi_i = \frac{v_0^{2/3}/s_0}{v_i^{2/3}/s_i} = \frac{(4\pi/3)^{2/3}/4\pi}{v_i^{2/3}/s_i}$$

where v_i and s_i are, respectively, the volume and whole surface of the Voronoi polyhedron i, while v_0 and s_0 are, respectively, the corresponding values of a sphere. The minimum value of ξ is unity, which corresponds to a sphere, and the more distorted the Voronoi polyhedron, the larger the value ξ. The average $\langle\xi\rangle_i$ denotes the local distortion in the whole volume.

4 **Local symmetry parameter**
By modifying the bond orientational parameters originally proposed by Steinhardt et al. [7], we have introduced [1] the local symmetry parameter $|\langle\hat{w}_6\rangle_i|$, which is the average value of a parameter describing the local symmetry around each atom. We have shown [1] that the larger the degree of amorphousness. the larger this local symmetry parameter. In fact, this parameter takes the largest value when the local symmetry around an atom is icosahedral.

5 **Voronoi face parameters**
Another useful quantity to express the degree of amorphousness is the Voronoi face parameter n_α which is defined by the ratio of the number of α–edged faces in the system to the total number of the Voronoi faces. We have also shown [1] that the larger the degree of amorphousness, the larger the value n_5 and the smaller the value n_4 and n_6.

With a view to understanding the relaxation of atomic configurations in disordered metastable structures, we study the behaviours of these parameters during the annealing processes. Special emphasis is laid on the investigation of the ways in which the relaxation in the mentioned systems is related to the fluctuations of atomic structures. In fact, it will be shown that the relaxation of the structure in a glass (or in other words in an amorphous solid) is is fairly distinct from that in a supercooled liquid.

Figure 6 shows several microscopic structure parameters as functions of the time steps in the annealing process at T=0.46. Soon after the onset of the annealing, the volume V, the distortion parameter $\langle\xi\rangle_i$, the local symmetry parameter $|\langle\hat{w}_6\rangle_i|$ and the Voronoi face parameter n_5 for pentagons begin to decrease at the same time which indicates the appearance of the crystalline nuclei.

The system is considered to be in the first stage of crystallization up to 3000 time steps or so, where the rearrangement of atoms is such that the local symmetry around some atoms changes from icosahedral to crystalline type without causing appreciable volume reduction. Since both δv and $\langle\xi\rangle_i$

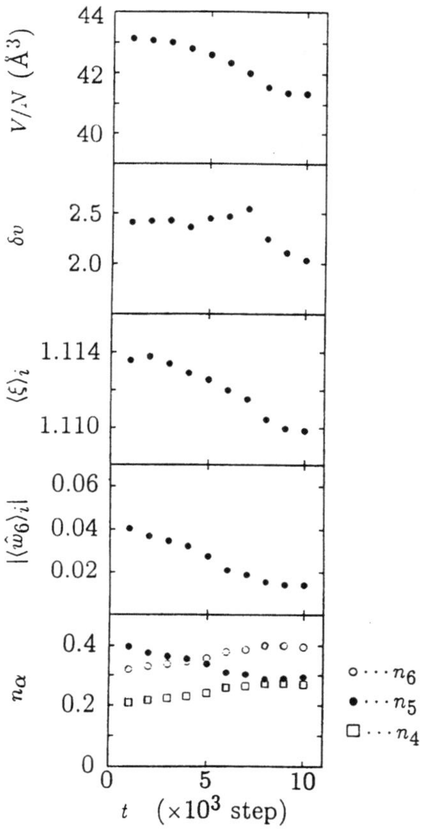

Fig. 6. Microscopic structure parameters versus annealing time at T=0.46. V/N denotes the average of the volumes of the Voronoi polyhedra, δv the fluctuation parameter, ξ the distortion parameter, $|\langle \hat{w}_6 \rangle_i|$ the local symmetry parameter, and n_α the Voronoi face parameters.

remain practically unchanged at this stage, the rearrangement of atoms is still local.

The rearrangement of atoms becomes more extensive at the second stage of crystallization, which corresponds to the time steps between t=3000 and t=5000. All the parameters expressed by solid circles in the figure, except for the fluctuation parameter δv, keep decreasing continuously at this stage, while n_6 for hexagons, denoted by open circles, and n_4 for quadrilaterals, denoted by open squares, are increasing. These behaviours of the parameters suggest that the system is in the coexistence state composed of both crystalline and liquidlike regions.

The transition from this second stage to the final stage of the crystallization seems to take place between t=5000 and t=6000, during which period the parameters $|\langle \hat{w}_6 \rangle_i|$ and n_5, reflecting the abundance of atoms with the icosahedral symmetry, decrease noticeably, while the fluctuation parameter δv starts increasing.

As implied from the projections at t=5000 in Fig. 4, where the crystalline clusters have grown to considerable sizes, a rather drastic rearrangement of the atomic configuration is required before the system achieves a single–crystal phase. This kind of the drastic rearrangement can be realized if the packing of atoms is not too tight to prohibit local occurrence of a large fluctuation in the atomic distribution.

In the case of annealing a supercooled liquid, there is still enough room for the large fluctuation to take place, as clearly shown in the fluctuation parameter δv at t=7000. This large fluctuation makes possible large–scale

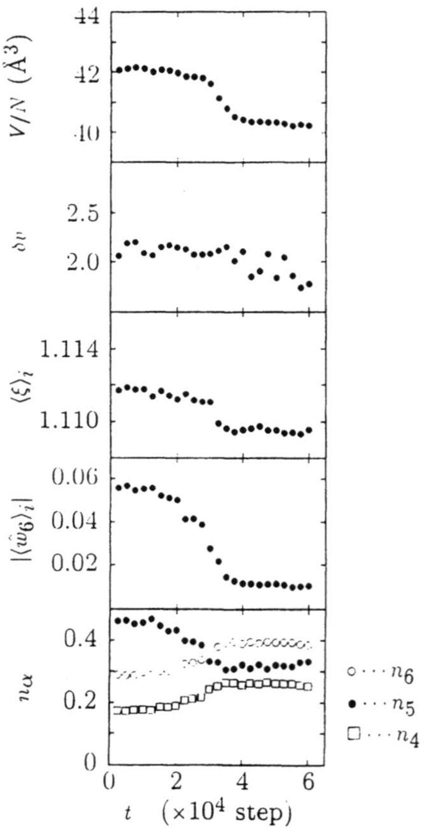

Fig. 7. Microscopic structure parameters versus annealing time at T=0.34. V/N denotes the average of the volumes of the Voronoi polyhedra, δv the fluctuation parameter, ξ the distortion parameter, $|\langle \hat{w}_6 \rangle_i|$ the local symmetry parameter, and n_α the Voronoi face parameters.

rearrangement of atomic positions, which is followed by substantial crystallization. After t=8000, the parameters V, $|\langle\hat{w}_6\rangle_i|$ and the Voronoi face parameters(n_4, n_5, and n_6) are all nearly constant, which suggests that the atomic configuration has attained crystalline symmetry after the large fluctuation at t=7000. On the other hand, the fluctuation parameter δv itself decreases further after t=8000, thus indicating that the deviations from the regular lattice points become smaller with time.

It is interesting to study the time development of various microscopic parameters in the process of annealing a glass at T=0.34 (shown in Fig. 7) in comparison with those in the process of annealing a liquid at T=0.46 (shown in Fig. 6). In particular, note the differences in time scales as described in the abscissa in Figs. 6 and 7. The relaxation time in the latter case in longer by one order of magnitude than that in the former.

The parameters $|\langle\hat{w}_6\rangle_i|$ and n_5 for symmetry start decreasing after t=20,000 when the reductions in V and $\langle\xi\rangle_i$ are not appreciable. This is different from the situation observed at the first stage of annealing at T=0.46, where all four parameters start decreasing at the same time. The difference is considered to come from the fact that, in the glass, the atoms are packed so tightly that the clusters with crystalline structures are created as a consequence of local rearrangement of atoms with icosahedral symmetry.

An essential contraction of the system starts after t=30,000, which is accompanied by the rather discontinuous reductions in $\langle\xi\rangle_i$, $|\langle\hat{w}_6\rangle_i|$ and n_5. These results suggest the progress of the crystallization. A most remarkable point, however, is the behaviour of the fluctuation parameter δv, which increases slightly toward t=30,000 but cannot change appreciably after that time step. This is again ascribed to the fact that the close packing of the atoms in the glass makes it difficult for an extensive adjustment to develop.

5. Distribution of the Volumes of the Voronoi Polyhedra

The assertions in the preceding section are reinforced by the analysis concerning the distributions $p(v_i)$ of the volumes of the Voronoi polyhedra during the annealing process.

Figure 8 is a plot of $p(v_i)$ for the annealing of a supercooled liquid at the same time steps as given in Fig. 4. In each figure, the solid curve represents $p(v_i)$. which is fitted through the least–square method by the sum of two Gaussian curves, described by the dot–dash and broken curves. Curve fitting by the Gaussian distributions gives an overall picture of the coexistence of ordered and disordered regions. The Gaussian curve at lower v_i values is expected to correspond to the ordered regions, while the curve at higher v_i values is regarded as expressing the disordered regions.

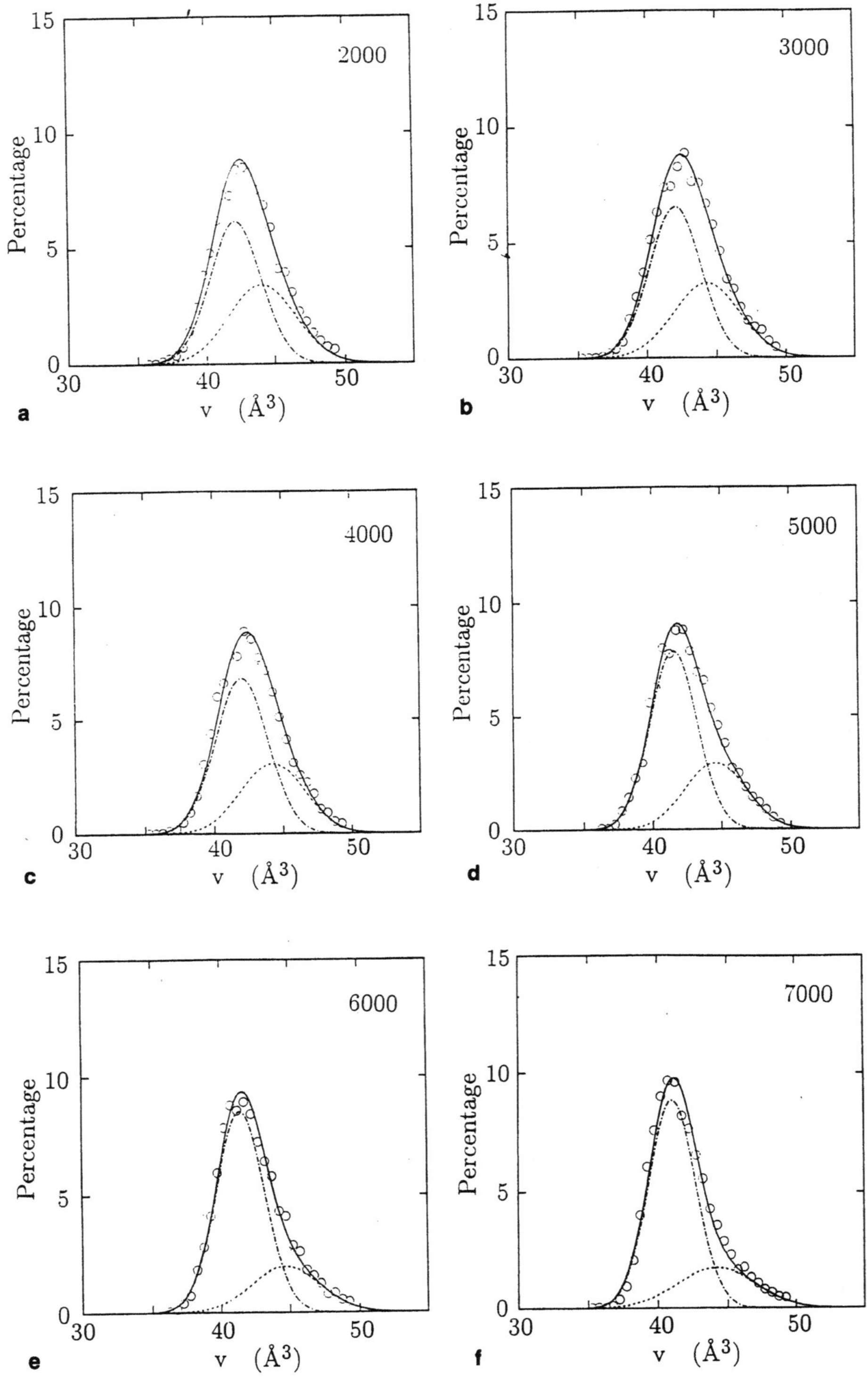

2000
3000
4000
5000
6000
7000
Percentage
0
5
10
15
30
40
50
v (Å^3)
a
b
c
d
e
f

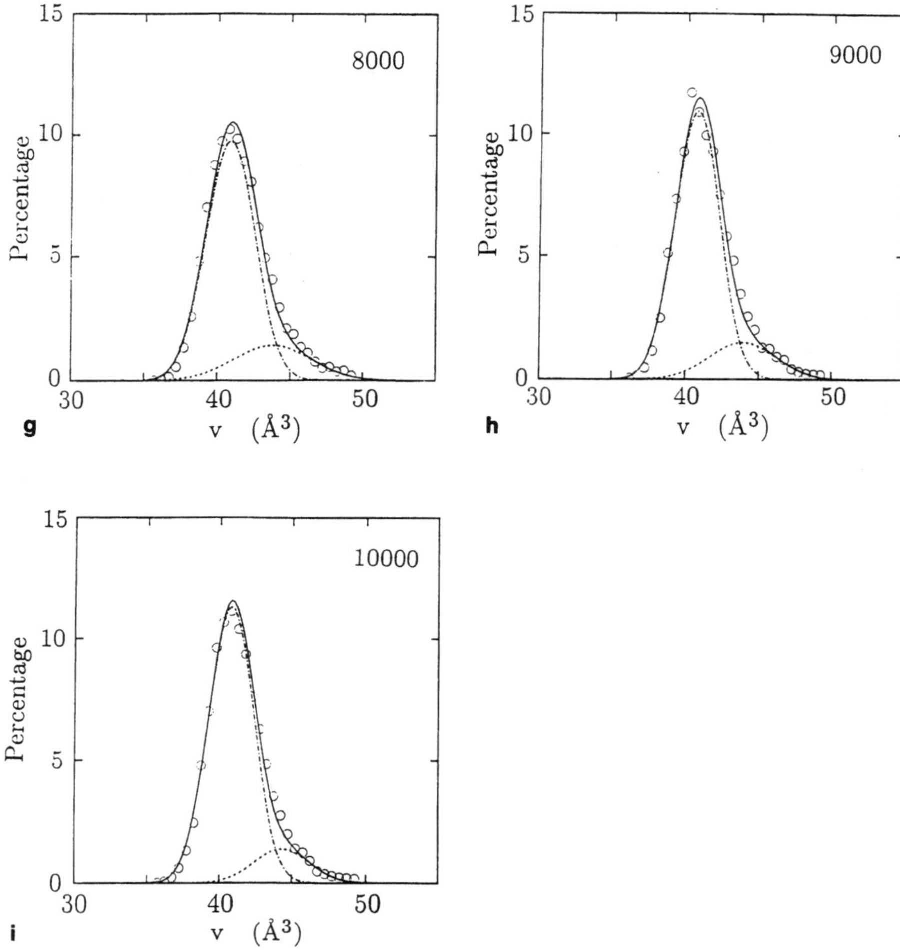

Fig. 8. Distribution of the volumes of the Voronoi polyhedra at every 1000 time steps between t=2,000 and t=10,000 on annealing a supercooled liquid at T=0.46.

According to the figures, there exist the first stage of the crystallization, the coexistence stage, a large fluctuation at 7000 time steps, and substantial crystallization at the final stage. The situation is more clearly demonstrated in Fig. 9 by the time–step dependencies of the mean values and standard deviations of the two Gaussian distributions. The Voronoi volumes of the atoms in the lower–volume group decrease gradually, reflecting the growth of the crystalline clusters as time proceeds. The standard deviation σ_1 of this lower–volume group also decreases monotonically, while the standard deviation σ_2 of the higher–volume group has a distinct maximum at t=7000 at which the fluctuation parameter δv increases markedly. These facts are consistent with each other, lending support to the argument that large–scale rearrangement of atoms is necessary to achieve a single–phase crystallization.

The distributions $p(v_i)$ for glass annealing at T=0.34 are presented in Fig. 10 for four time steps between t=25,000 and t=40,000. In this case, it is rather difficult to fit the whole curve by using the sum of two Gaussian curves. A detailed analysis indicates that $p(v_i)$ in this case of annealing a glass is decomposed into (1) a Gaussian distribution at lower Voronoi volumes, denoted by the dot–dash curve in each figure, and (2) the higher–volume region described by solid squares, the latter having a longer tail.

The difference in the characteristic features of the distributions $p(v_i)$ for a supercooled liquid and a glass is probably due to the fact that in a supercooled liquid a clear–cut distinction is possible between the coexisting ordered (crystalline) and disordered (liquidlike) regions, whereas this distinction cannot occur in a glass because the ordered (crystalline) and disordered (amorphous) regions have more or less similar densities.

In this way, microscopic structure analysis clarifies the distinction between the relaxation of a supercooled liquid and of a glass.

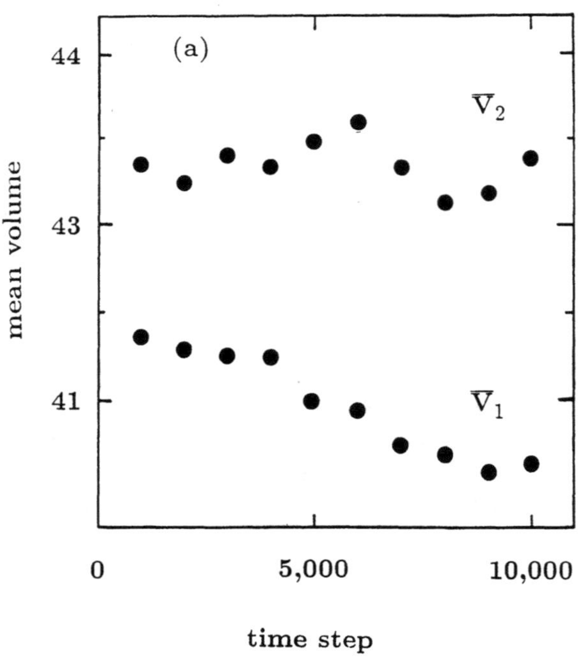

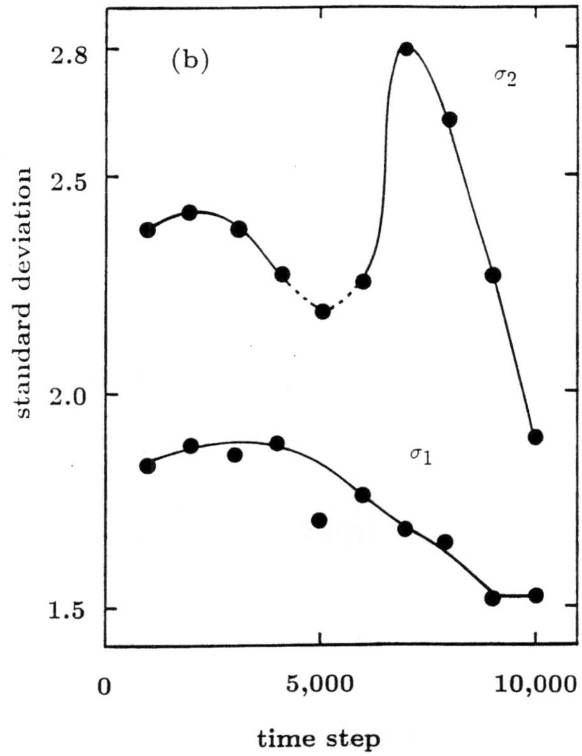

Fig. 9. The means v_1 and v_2 and the standard deviations σ_1 and σ_2 of the two Gaussian curves fitted to $p(v_i)$ during the annealing process at T=0.46.

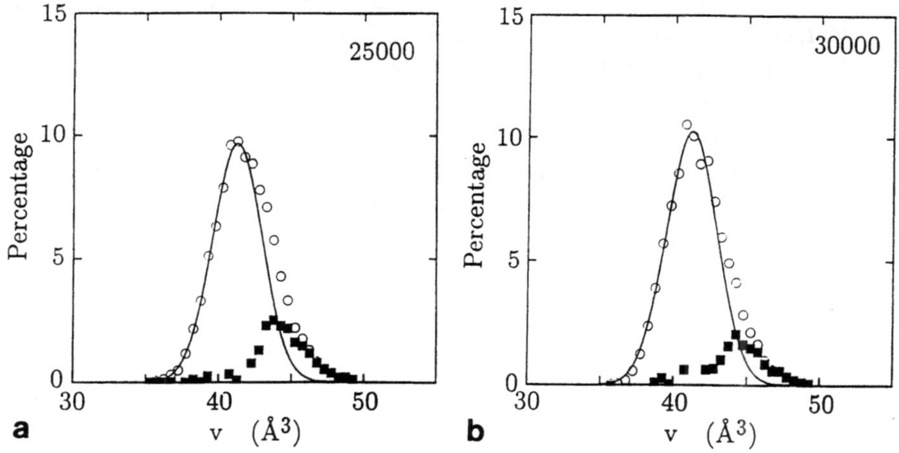

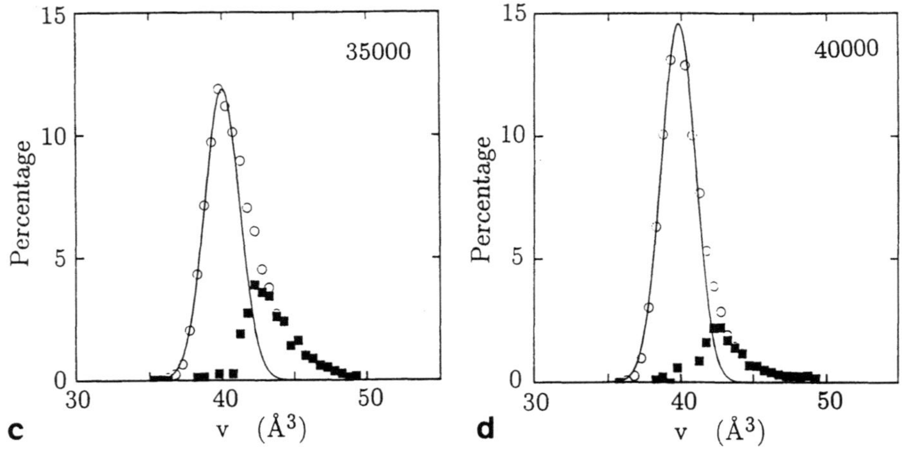

Fig. 10. Distribution of the volumes of the Voronoi polyhedra at every 5000 time steps between t=25,000 and t=40,000 on annealing a glass at T=0.34.

Acknowledgement

The authors are indebted to Mr. T. Yamaguchi for his help in the preparation of the manuscript.

References

1. F. Yonezawa: in *"Solid State Physics"*, vol. 45 (H. Ehrenreich and D. Turnbull, eds.), Academic Press, 1990.
2. M. Kimura and F. Yonezawa: in *"Topological Disorder in Condensed Matter"* (F. Yonezawa, ed.), Springer, 1983.
3. F. Yonezawa, S. Nosé and S. Sakamoto: Neue Folge **156**, 77 (1988).
4. W. Damgaard Kristensen: J. Non–Cryst. Solids **21**, 303 (1976).
5. H. C. Andersen: J. Chem. Phys. **72**(4), 2384 (1980).
6. J. L. Finney: Proc. Roy. Soc. **A 319**, 479 (1970).
7. P. Steinhardt, D. R. Nelson, and M. Ronchetti: Phys. Rev. **28**, 784 (1983).

Part III

Oxides and Multi-component Materials

Molecular Dynamics Simulation of Amorphous Silica

R.G. Della Valle and H.C. Andersen*

Department of Chemistry, Stanford University, Stanford, CA 94305, USA
*Permanent address: Università di Venezia, Dipartimento di Chimica-Fisica, Calle Larga S. Marta, Dorsoduro 2137, I-30123 Venezia, Italy

Abstract. We have performed molecular dynamics computer simulations of the cooling of liquid silica to form an amorphous solid (glassy) phase using the SiO_2 potential of Tsuneyuki, Tsukuda, Aoki, and Matsui, which they had found to be consistent with the structure of several crystalline phases of silica. The purpose of our present work was to provide further tests of the potential and to see what insight could be gained about the structure of amorphous silica and the nature of the glass transition in this material. In our simulations, the atoms formed an almost perfect random tetrahedral network, and on subsequent annealing the material formed a perfect network, with each silicon having four oxygen neighbors and each oxygen having two silicon neighbors. The simulated material at low temperatures has a structure that is in close agreement with experimental neutron scattering measurements on amorphous silica. Several properties of the low temperature material were calculated and compared with experiment, and the nature of the 'inherent structures' of the material as a function of temperature was investigated. The potential of Tsuneyuki *et al.* provides an excellent model for amorphous silica as well as crystalline silica.

1. Introduction

A number of molecular dynamics simulations of silica glass have been performed using effective interatomic potentials [1-10]. The potentials developed for this purpose typically have two body interactions including Coulombic and short range repulsive parts, and some of them include three-body contributions. Some of the parameters of the potential are typically obtained by using experimental information on silica glass. The existence of several crystalline forms of silica provides an opportunity for subjecting effective potential models to stringent tests. An ideal effective potential should be capable of describing all the condensed phases of silica, both crystalline and amorphous.

Recently Tsuneyuki *et al.* [11] have developed a two-body potential that was fit to *ab initio* calculations on tetrahedral SiO_4^{4-} clusters. They found that this model reproduces the crystallographic structures, energies, and compressibilities for all the more common silica polymorphs. We [12] tested the potential by calculating phonon frequencies for the silica polymorphs and found remarkably good agreement with experiment. The potential has been applied with success to the study of pressure induced [13] and thermal induced [14] phase transitions in crystalline silica. The model has also been found suitable for an MD simulation of liquid silica under pressure [15]. The bond angles in the polymorphs are reproduced quite precisely [11], thus suggesting that this model might be able to reproduce the angles also for the glass.

Because of the great interest in glassy silica and the success of the potential of Tsuneyki *et al.* in accounting for the properties of the crystalline phases of silica, we decided to perform molecular dynamics studies of the liquid and glass using this potential.

2. Methods and Calculations

The potential model of Tsuneyuki *et al.* [11] employs pairwise additive interatomic potentials of the form $V_{ij}(r) = q_i q_j / r + A_{ij}\exp(-B_{ij}r) - C_{ij}/r^6$, which consists of a Coulomb interaction, Born-Mayer repulsion, and an attractive interaction. No three-body or many-body interaction is included. (Due to the divergence of the r^{-6} term, the O-O, Si-O and Si-Si potentials become unphysically attractive at short distances. A discussion of this point and a plot of the potential functions are available in reference [15]. In our simulations, which have been started from a totally random configuration, we have found that such short distances were sampled only during the initial non-equilibrium transient in the first few hundred time steps. In this initial transient we have replaced the short distance attractive part of the force with a constant repulsive force.)

The long ranged Coulomb interaction were evaluated by Ewald's method [16], which is required to ensure that the correct equilibrium structure of the liquid is obtained [17]. A charge distribution width $w = L/5.714$ [16] has been used, while the R-space and K-space sums have been extended up to 10 Å and $5 \times 2\pi/L$ respectively. L is the edge of the simulation cell. These values have been chosen to reduce the series truncation error in the potential energy to less than one part in 10^5 while conserving computer time.

We have simulated a sample of pure SiO_2 consisting of 500 O atoms and 250 Si atoms. All simulations were at constant volume, for a density of 2.2 g/cm^3 [18]. The velocity form of the Verlet algorithm [19,20] was used to integrate the equations of motion. The simulation was started with totally random positions and a Boltzmann distribution of velocities appropriate to 4000 K, inside a cubic cell with periodic boundary conditions. During the first few hundred simulation steps we have used short time steps (of order 0.01 fs) to avoid numerical instabilities in the integration of the equations, and the velocities were rescaled at every time step to obtain the desired kinetic energy. After the initial transient, the velocity rescaling was discontinued, and in the subsequent calculations a time step of 1 fs was used. This is about 1/25 of the shortest vibrational period computed for the crystalline silica polymorphs with the same potential [12].

This sample was equilibrated, for more than 20000 fs, in contact with a heat bath at 4000 K. This temperature is well above the experimental melting point of all the silica polymorphs [21]. The contact with the heat bath was simulated using the stochastic collision method [22] with an average of 1 collision/fs for the whole sample.

The temperature of the bath was then continuously lowered from 4000 to 300 K, with a cooling rate of 0.025 K/fs between 4000 and 1400 K (104000 fs), and of 0.25 K/fs below 1400 K (4400 fs). On the time scale of our simulations no rearrangement of the network structure takes place at 1400 K, which is below the experimental glass transition temperature. Thus the faster cooling rate at lower temperature saves a considerable amount of computer time without affecting the final structure. The bath was maintained at 300 K for an additional 20000 fs, to allow the sample to reach thermal equilibrium. The approach to the equilibrium, which was sluggish, was monitored by following the stabilization of the kinetic and potential energies.

At various times during the cooling, the dynamical state was recorded. Subsequently, some of these states were used as the starting points for long (15,000 fs or more) constant energy simulations that allowed the structure to relax and anneal. In this paper we report the properties of three different types of states: 1. the 300K state that was finally achieved after cooling and equilibration; 2. the instantaneous states that the system passed through during the cooling run; and 3. the states the system passed through during the constant energy annealing runs.

In order to obtain increased understanding of the structure of the simulated systems, we employed the 'inherent structure' method of Stillinger and Weber [23]. The inherent structure of a system in a particular dynamical state is defined as the structure at the local minimum on the many particle potential energy surface that is obtained by steepest descents minimization of the potential energy starting from the dynamical state. The inherent structure can then be regarded as the local potential energy minimum about which the system is oscillating in the dynamical state. We used a modified MD algorithm to find the potential minima. In the modified algorithm, the system is started at the initial configuration (with the velocities initially set equal to zero) and then allowed to move forward in time according to the equations of motion. After each time step, for each particle the dot product of the velocity of the particle and the force on the particle is calculated. The velocity of the particle is then replaced by the projection of the velocity on the force (if the dot product is positive) or by zero (if the dot product is negative). The system is followed until the kinetic energy becomes sufficiently low. With this method the system is never able to surmount a potential energy barrier and rapidly falls towards the local minimum around which it found itself at the beginning of the quench. For silica, between 500 and 1500 steps (with a time step of 3 fs) are usually enough to reach a point at which the potential energy no longer decreases at an appreciable rate. In practice, we stopped the calculation when the average kinetic energy corresponded to some low temperature, either 0.01 K or 1 K depending on the type of system considered.

3. Results

3.1 Low temperature material

The sample obtained from the cooling of the liquid and annealing at 300 K has a pressure of -2.2 ± 0.2 GPa at the density of 2.2 g/cm^3. By scaling the volume and coordinates in small steps in a series of short simulation runs, we have found that with a 5% reduction of the volume the pressure becomes zero. Most of the results of the simulation for the sample at 2.2 gm/cm^3 density may be considered as valid also for zero pressure, since the differences in density (5%) and energy (0.5 kcal/mol) for the two pressures are small. However it must be kept in mind that the interatomic distances obtained in the simulation are likely to be too large by 1–2% because of the 5% error in density.

As expected, the average molar energy of the glass (−1224.2 kcal/mol at zero pressure) is above the energy computed for the crystalline silica polymorphs with the same potential model [11]. The difference between the computed energy of glass and of cristobalite [11], the high temperature polymorph, is 6.8 kcal/mol, which should be compared to the nearest experimental equivalent, the enthalpy difference between glass and cristobalite, which has a value of 1.5 kcal/mol [24].

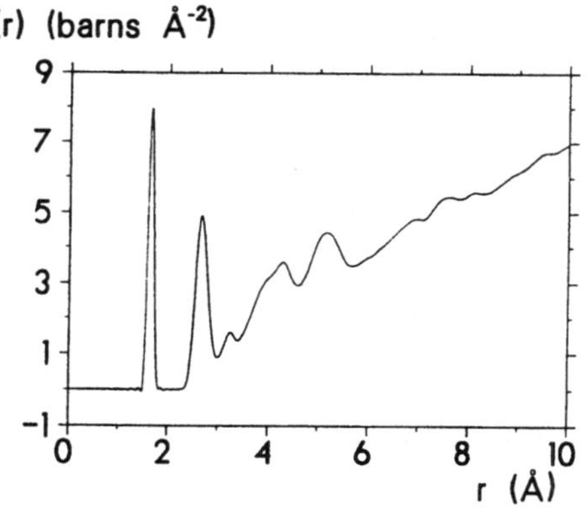

Figure 1. The total correlation function for neutron scattering, $T(r)$, calculated from the structure of the simulated low temperature glassy material. The notation is that of Grimley *et al.* [28]. These results should be compared with those of Grimley *et al.* and of Wright *et al.* [27].

The bulk modulus, $B = (1/V)(\partial p/\partial V)_T$ for the simulated system is 41.8 GPa, which is close to the experimental value, 36.9 GPa [25].

We have computed the pair correlation functions for the atomic species. The total correlation function for neutron scattering, $T(r)$, was then obtained by adding the three pair correlation functions with the appropriate weights [26] and convoluting the sum with the same experimental resolution function that is used in the neutron scattering experiments [27,28]. The result is shown in Figure 1, which should be compared with the data in the papers describing the experimental results. The agreement between the computed and experimental $T(r)$, in the position, shape and area of the peaks at short and intermediate distances is very satisfactory. The structure of the simulated glass seems to reproduce very closely that of the real material.

3.2 Heat capacity and the glass transition

At low temperatures the constant volume heat capacity, C_V, has the classical value for an harmonic solid of SiO_2, $C_V = 9k_B = 74.83$ J/mol K. The heat capacity is larger at higher temperatures, due to the structural rearrangements that take place during the cooling, and to the vibrational anharmonicity. In the temperature range of 1500 to 2500 K, the computed C_V is 81.06 J/mol K, which agrees with the experimental value measured for the liquid in this temperature range [29], $C_p = 77.75 \pm$

3.88 J/mol K. The difference between constant pressure and constant volume heat capacities, C_p and C_V, is expected to be small, since the thermal expansivity of silica is small.

For most glass forming materials the experimental heat capacity measured during a cooling process drops at the glass transition from a rather large "liquid-like" value to a "solid-like" value near to that appropriate to an ideal harmonic solid [30,31]. In our simulation results, no perceptible sudden change in the slope of C_V may be seen. This result is in agreement with the experimental fact that SiO_2 is one of the few materials that do not show an observable discontinuity in the heat capacity at the glass transition [31]. The experimental and computed thermodynamic properties of liquid SiO_2 merge continuously with those of the glass. The identification of the glass transition temperature in our simulation must be based on different criterion than a drop in the heat capacity. In the following section we discuss how to use the inherent structures to identify the transition temperature.

3.3 Inherent structures and the glass transition

As discussed above, we have calculated the inherent structure for various states that the system achieved during the cooling run. The potential energy of the inherent structure as a function of the sample temperature is shown in figure 2. As the temperature is lowered from 4000 K to about

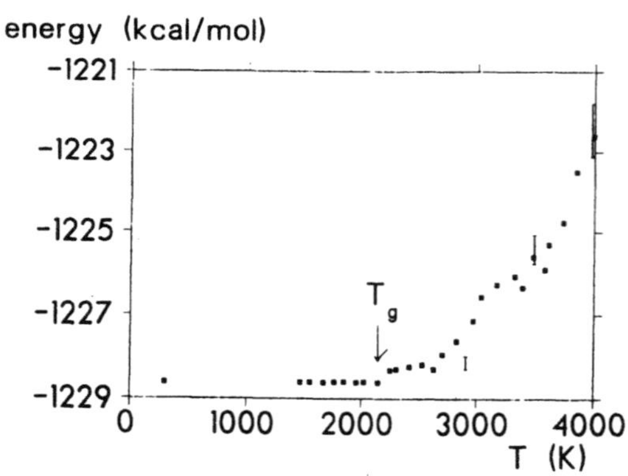

Figure 2. The potential energy of the inherent structure as a function of temperature during the cooling run. See the text for the meaning of the vertical line segments at high temperatures. Also see the text for the basis of our conclusion that the glass transition occurs in our simulation at a temperature of 2200 K.

2200 K, the energy of the inherent structures drops, but below 2200 K it is essentially constant. We also calculated the inherent structure for various states that the system achieved along the annealing runs. At each temperature, the system samples inherent structures with a variety of energies. The mean energy of the inherent structures plus and minus one standard deviation are given in Figure 2 for three high temperatures. They are represented by the vertical line segments. Below about 2200 K, the system stays in one inherent structure during the entire annealing run.

The interpretation of these results is straightforward. At high temperatures the system succeeds in hopping from one local minimum on its potential surface to another and it sees a variety of energies of minima. As the temperature is lowered the system spends time in the vicinity of progressively lower and lower energy minima and the range of energies that it can explore narrows. Below about 2000 K the system is incapable of overcoming the barriers between minima. We can identify this with the absence of what in the glass literature is referred to as 'structural relaxation'. The point at which structural relaxation no longer can take place corresponds to the glass transition. In our simulation, the energy of the inherent structure ceases to change, and hence the glass transition occurs, at about 2200 K. Moreover, the range of energies explored is essentially zero below this temperature. On the molecular dynamics time scale, the system is trapped in a single local minimum on the potential energy surface at temperatures of 2200 K and below. As expected [30], this is higher than the experimental glass transition temperature of 1446 K [32].

3.4 Activation energy for transport

We have calculated the self-diffusion coefficients of the silicon and oxygen atoms as a function of temperature by monitoring the mean squared displacement of atoms as a function of time interval in constant energy simulations. The runs were about 15000 fs in length. Above 3500 K a thousand fs suffice to reach a rather complete equilibration, while below 3000 K the equilibration cannot be reached in practically obtainable computer times.

The activation energy (in temperature units) for the diffusion coefficients, estimated from the last 10000 fs of the runs at higher temperatures is 36700 $\pm$ 2100 K. This value is to be compared with the experimental activation energy for the viscosity, which is in the 63000-75000 K range [31,33]. It is expected that diffusion coefficients and viscosity coefficients should have the same temperature dependence, since both relaxation processes are probably limited by the same temperature dependent structural relaxation process. Thus we have a significant discrepancy between the model system and real silica. The discrepancy is mainly to be attributed to insufficiencies in the potential model.

4. Discussion of the structure of the glass

We have performed an analysis of the structure of the material as a function of temperature. The details will be published elsewhere. We shall limit the present discussion to a few comments.

The structure of the low temperature glass corresponds to the usual network model for SiO_2, namely that it consists of SiO_4 tetrahedra that share corners. Below the glass transition the sample is a random tetrahedral network essentially free of coordination defects. All of the 250 Si atoms have four O atom neighbors, and 498 of the 500 O atoms have two Si neighbors. The two remaining defects are two oxygen atoms with coordination numbers of one and three respectively. All of 500 pairs of connected SiO_4 tetrahedra are corner-sharing, except one that is edge-sharing. A different sample, annealed for 6000 fs just above T_g and then cooled below, contained no coordination defects and no edge-sharing tetrahedra.

Above the glass transition we find a number of coordination defects that grows with the temperature of the liquid. The percentage of atoms involved in defects is always small, less than 4% at 4000 K. Due to the strong repulsive forces between the silicon atoms, the number of edge-sharing pairs of polyhedra is small at all temperatures (5 or less, against about 500 corner-sharing pairs), and no pair of face-sharing polyhedra is ever found.

The other major temperature dependence observed in the structure is in the statistics of the distribution of rings. As the temperature is lowered, the number of small rings (containing two or three silicon atoms) decreases significantly. The numbers of rings of other sizes stays approximately constant with temperature.

There may be other aspects of the structure that change with temperature that we have not yet discerned.

5. Concluding Remarks

These calculations show that the model potential of Tsuneyuki *et al.* gives a remarkably good representation of the properties of liquid and amorphous silica. The form of the potential was motivated by *ab initio* quantum mechanical calculations, and the choice of parameters was guided by using experimental information on the compressibility and structure of α-quartz. However, *no* liquid state or glassy state data was apparently used in the construction of the potential. All the results in this paper can then be regarded as theoretical predictions of the properties of liquid and amorphous silica. The generally accepted random tetrahedral network structure formed spontaneously in the simulations, and many of the structural and thermodynamic properties of the simulated material are in good agreement with the measured properties of amorphous silica.

It is also remarkable that a defect-free random tetrahedral network structure could be spontaneously formed by cooling a high temperature random configuration on a molecular dynamics time scale. The reason for this is apparently that the barrier to migration and mutual annhiliation of coordination defects of the random tetrahedral network is low enough in this model so that such migration takes place at temperatures as low as about 2200 K. (In fact, there is some indication, from the self-diffusion calculations discussed above, that the activation energies for processes of structural change are somewhat too small for the model as compared with experiment.)

The potential model is good enough that the results of these simulations should provide some valid insights into the nature of the glass transition, the structure, and the structural relaxation of silica. The results of more detailed analysis of our calculations will be published elsewhere. Moreover, this potential model may provide the basis for development of potentials for the modeling of crystalline, liquid, and glassy silicates of geological interest.

6. Acknowledgements

This work was supported by the National Science Foundation (grants CHE84-10701 and CHE89-18841). The computations made use of facilities provided by NSF grant CHE88-21737. We are grateful to the Italian C.N.R. for financial support of one of us (RGDV).

7. References

1. L. V. Woodcock, C. A. Angell, and P. Cheeseman, J. Chem. Phys. **65**, 1565 (1976).

2. C. A. Angell, P. A. Cheeseman, and C. C. Phifer, Mat. Res. Soc. Symp. Proc. **63**, 85 (1985).

3. T. F. Soules, J. Chem. Phys. **71**, 4570 (1979) ; **73**, 4032 (1980).
4. S. H. Garofalini, J. Chem. Phys. **76**, 3189 (1982).
5. N. V. Doan, Phil. Mag. **A49**, 683 (1984).
6. S. K. Mitra, M. Amini, D. Fincham, and R. W. Hockney, Phil. Mag. **43**, 365 (1981).
7. S. K. Mitra, Phil. Mag. **45**, 529 (1982) ; **47**, L63 (1983).
8. S. K. Mitra and J. M. Parker, Phys. Chem. Glasses **25**, 95 (1984).
9. B. P. Feuston and S. H. Garofalini, J. Chem. Phys. **89**, 5818 (1988).
10. B. Vessal, M. Amini, D. Fincham, and C. R. A. Catlow, Phil. Mag. **B60**, 753 (1989).
11. S. Tsuneyuki, M. Tsukada, A. Aoki, and Y Matsui, Phys. Rev. Lett. **61**, 869 (1988).
12. R. G. Della Valle and H. C. Andersen, (submitted to the Journal of Chemical Physics).
13. S. Tsuneyuki, Y. Matsui, H. Aoki, and M. Tsukada, Nature **339**, 209 (1989).
14. S. Tsuneyuki, H. Aoki, and M. Tsukada, Phys. Rev. Lett. **64**, 776 (1990).
15. J. R. Rustad, D. A. Yuen, and F. J. Spera, Phys. Rev. **A42**, 2081 (1990).
16. L. V. Woodcock and K. Singer, Trans.Faraday Soc. **67**, 12 (1971).
17. R. L. Erikson and C. J. Hostetler, J. Geochim. Cosmochim. Acta, **51**, 1209 (1987).
18. R. L. Mozzi and B. E. Warren, J. Appl. Cryst. **2**, 164 (1969).
19. W. C. Swope, H. C. Andersen, P. H. Behrens, and K. R. Wilson, J. Chem. Phys. **76**, 637 (1982).
20. M. P. Allen and D. J. Tildesley, "Computer Simulation of Liquids", Clarendon Press, Oxford (1987).
21. JANAF Thermodynamical Tables, **14**, 1673 (1985).
22. H. C. Andersen, J. Chem. Phys. **72**, 2384 (1980).
23. F. H. Stillinger and T. A. Weber, Phys. Rev. **A25**, 978 (1982).
24. J. L. Holm, O. J. Kleppa and E. F. Westrum, J. Geochim. Cosmochim. Acta, **31**, 2289 (1967).
25. K. Kondo, S. Iio, A. Sawaoka, J. Appl. Phys. **52**, 2826 (1981).
26. G. E. Bacon, Acta Cryst. **A28**, 357 (1972).
27. A. C. Wright, A. G. Clare, D. I. Grimley, R. N. Sinclair, J. Non-Cryst. Solids **112**, 33 (1989).
28. D. I. Grimley, A. C. Wright, R. N. Sinclair, J. Non-Cryst. Solids, **119**, 49 (1990).
29. N. D. Stout and A. J. Piwinski, High-Temperature Science, **15**, 275 (1982).
30. C. A. Angell, J. H. R. Clarke and L. V. Woodcock, Adv. Chem. Phys. **48**, 397 (1981), and references therein.

31. S. Brawer, "Relaxation in Viscous Liquids and Glasses", Am. Ceramic. Soc., Columbus, Ohio (1985).
32. C. A. Angell, J. Chem. Phys. Solids **49**, 863 (1988).
33. G. Hetherington, K. H. Jack and J. C. Kennedy, Phys. Chem. Glasses **5**, 130 (1964).

Pressure-Induced Structural Transformations and Diffusion Mechanism in Silica

S. Tsuneyuki

Department of Physics, University of Tokyo, 7-3-1 Hongo, Bunkyo-ku, Tokyo 113, Japan

Abstract Melts of many silicates are known to have an anomalous transport property that the viscosity drastically decreases with increasing pressure. Here we propose a mechanism for the anomalous diffusion process in liquid silica and silicates on the basis of molecular dynamics (MD) results for pressure-induced structural transformations in crystalline silica. The conjecture is tested with an MD simulation for silica melt.

1 Introduction

Transport properties of silicate melts have been intensively studied because of geological interest, since they are closely related with the igneous processes in the earth. A conspicuous and anomalous feature is that the viscosity in the liquid phase of many silicates such as jadeite ($NaAlSi_2O_6$) drastically decreases with increasing pressure [1,2] just as in the case of water and aqueous solutions. Microscopically this implies that the principal elements in the silicate melts, *i.e.*, oxygen and silicon (and aluminum) atoms diffuse more easily at higher pressures, as confirmed experimentally [3].

A characteristic feature common to the silicates that exhibit the anomalous diffusion property is that they have highly polymerized framework structures comprising corner-shared SiO_4 (or AlO_4) tetrahedra. In fact, when the framework is destroyed by a sufficient amount of alkali or alkaline earth metals as in the melt of diopside ($CaMgSi_2O_6$), the system no longer shows such an anomaly and the viscosity increases with pressure [4].

Molecular dynamics (MD) studies on the anomalous diffusion property were carried out by Angell et al. [5,6]. They have successfully shown that, in the melt of jadeite composition in which the SiO_4 (AlO_4) tetrahedra are well polymerized, the self-diffusion coefficients of oxygen, silicon and aluminum atoms have maxima at a pressure of 20-30 GPa, whereas the diffusion coefficient in the depolymerized melt of diopside monotonically decreases with pressure. On the mechanism of the anomalous diffusion process, they have pointed out the importance of five-fold coordinated silicon atoms, a fraction of which also has a maximum at the pressure corresponding to the maximum diffusivity.

Although their MD results show an qualitative agreement with the experimental results, and although we believe their insight into the diffusion is essentially relevant, the mechanism of the anomalous diffusion process is still not too clear because of ambiguities inherent in their simulation. The largest ambiguity concerns the interatomic potential, which qualitatively dominates the MD simulation of condensed matters. Angell et al. empirically determined the potential using only static data at low pressures. Inclusion of elements that destroy the framework further complicates the discussion.

In this context, here we re-investigate the mechanism of the anomalous diffusion process in silicate melt by a non-empirical MD simulation of silica (SiO_2). The viscosity of the framework GeO_2 has been experimentally shown [7] to have anomalous diffusion, and it is considered that the silica melt should also show an analogous anomaly. Silica, an end member of silicate compounds, has a complete framework structure, so that the material is an ideal starting point for theoretical analysis.

The pairwise interatomic potential used in this study was first derived from first-principles electronic structure calculations for small clusters. Although there have been several empirical approaches to obtaining pairwise interatomic potentials for silicate, non-empirical determination of the potential is obviously desirable, and we have in fact succeeded in simulating the structures and bulk moduli of all the known polymorphs of silica at room temperature [8]. The potential was then successfully used in the simulation of thermally induced phase transition of quartz [9], and also pressure-induced structural transformations with a slight refinement of the parameters [10]. Thus the validity of the interatomic potential used here have been examined in a number of ways.

In Section 2, we briefly review the MD results on the pressure-induced structural transformations in crystalline silica. A microscopic model for atomic diffusion, directly derived from the MD results, is depicted in Section 3, which leads us to the anomalous diffusion property of silicate melts. The conjecture is examined in Section 4 by an MD simulation of silica melt by controlling the potential parameters.

2 Pressure-Induced Structural Transformations at Room Temperature

In this section, we briefly review our MD simulation on the compression of crystalline silica at room temperature [10], which was motivated to investigate the microscopic mechanism of pressure-induced amorphization [11].

In the simulation, hydrostatic pressure was applied to four polymorphs of silica (low-quartz, low-cristobalite, coesite and stishovite) at room temperature with the constant-pressure algorithm due to Parrinello and Rahman [12] and the constant-temperature algorithm due to Nosé [13]. The equations of states obtained by the simulation are shown in Fig. 1 together with the experimental results in the inset [11].

The most important point is the stability of each polymorph. Low-quartz and coesite undergo structural transformations with discontinuous volume reduction at pressures of 25 to 30 GPa and 35 to 40 GPa, respectively. The critical pressures are in good agreement with those for

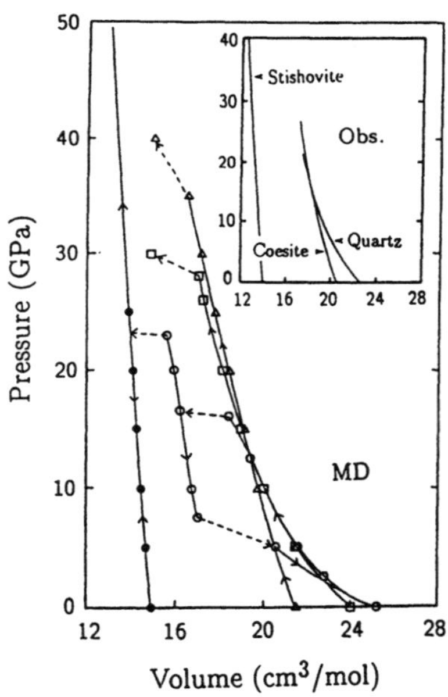

Figure 1: (a)Pressure-volume relation for low-quartz (□), coesite (△), stishovite (●) and low-cristobalite (○) obtained by MD [10]. Solid lines represent continuous change and broken lines discontinuous change of volume due to pressure-induced structural transformation. Hysteresis is indicated by arrows. Pressure-volume relation obtained by experiments is shown in the inset (from Hemley et al. [11])

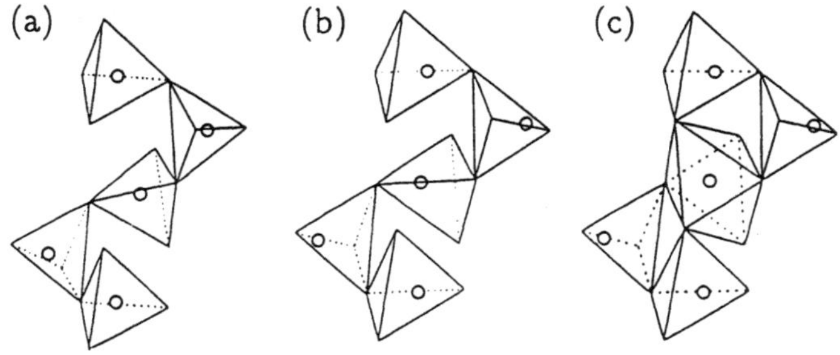

Figure 2: Mechanism of the pressure-induced structural transformation. (a) A framework of corner-shared SiO_4 tetrahedra in the low-cristobalite structure, where silicon atoms are shown with circles. (b) The framework structure deformed under pressure. (c) SiO_6 octahedron explicitly shown using the same atomic configuration as that in (b).

pressure-induced amorphization obtained experimentally. Though the detail of the final structures depends on the MD runs, they are amorphous for coesite and almost crystalline for low-quartz, both of which contain octahedrally coordinated silicon atoms. Similarly, low-cristobalite shows a structural transformation at about 16.5 GPa into a novel polymorph with space group *Cmcm* which contains both four-fold and six-fold coordinated silicon atoms. Further compression up to about 23 GPa leads to the second transformation into stishovite, in which all silicon atoms are octahedrally coordinated [14].

Interestingly, these structural transformations occur without atomic diffusion in the sense that no isolated atoms nor disordered states appear at the transition. The mechanism of these diffusionless transformations is illustrated in Fig. 2. Fig.2(a) represents the framework of corner-shared SiO_4 tetrahedra in the low-cristobalite structure. When the pressure is increased the framework structure is continuously deformed into a structure shown in Fig.2(b). This structure can be regarded as comprising a six-fold Si-O coordination if we examine the Si-O bond length: an SiO_6 octahedron is explicitly depicted in Fig.2(c) for the same atomic configuration as that in Fig.2(b). Thus the Si-O coordination number increases smoothly by neither breaking any Si-O bonds nor introducing global atomic diffusion. This mechanism for compression is essentially the same as that suggested by Stolper and Ahrens for amorphous silica [15].

From the mechanism above, it is naturally expected that the transformation from four-fold to six-fold coordination can be reversible. In fact, most high-pressure phases obtained here cannot be quenched upon decreasing the pressure, and become an amorphous phase, in which all silicon atoms are tetrahedrally coordinated. Moreover, the fact that the decompression gives rise to an amorphous rather than a crystalline phase means that the way of unfolding six-fold coordination into four-fold coordination is locally not unique.

3 A Model for Atomic Diffusion

On the basis of the MD results given above, we can make a microscopic model for atomic diffusion in silica. We have stressed that the obtained structural transformations are diffusionless. However, this is in a specified meaning of the word, in which a diffusive transition is defined as the one accompanied by the appearance of isolated atoms. Actually the diffusionless transition modes contribute to the mechanism of diffusion in the following sense.

When low-cristobalite is compressed into the *Cmcm* phase followed by a decompression to the normal pressure, the system becomes an amorphous phase, in which the oxygen coordinations are topologically rearranged from those in the starting cristobalite, although no isolated oxygens appear during the process. The coordination of silicon atoms only changes from four-fold to six-fold and then to four-fold again with different oxygen atoms as illustrated in Fig.3(a). If we envisage a process in a disordered phase in which the compressions and decompressions are repeated, this could result in a diffusion of atoms. Since local pressure is thought to fluctuate in a melt, we can expect such a diffusion in a liquid phase. Six-fold coordination is then envisaged as a local and transient state for diffusion. Note that the diffusion considered here does not accompany isolated atoms but is regarded as local changes of Si-O coordination numbers, which will require much less energy than that needed for stripping an oxygen atom from a four-fold coordinated silicon atom.

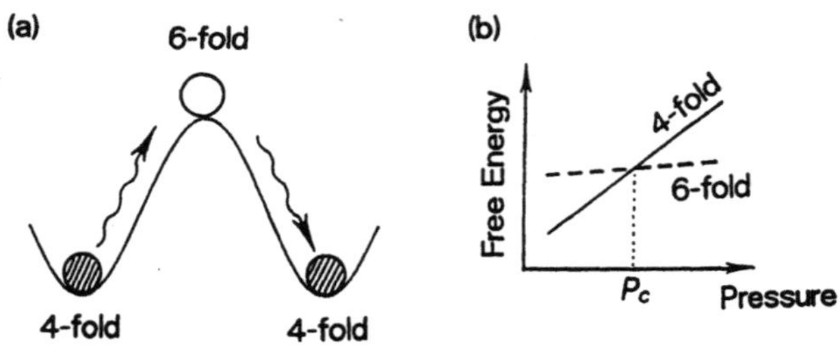

Figure 3: (a) Schematic view of atomic motion in configurational space, and (b) the pressure dependence of the free energy of the four-fold and six-fold coordinated systems.

Now the anomalous diffusion property can be explained naturally as follows. The crucial factor for atomic diffusion in the present model is the activation energy from four-fold coordination to six-fold coordination. As known from experiments, silicon atoms always have four-fold coordination at low pressures, while they have six-fold coordination only at high pressures, so that the free energies of the four-fold and six-fold coordinated states have different pressure dependences, as is schematically shown in Fig.3(b). Thus the activation energy decreases up to a certain critical pressure P_C, at which the two free energies coincide, and the diffusion coefficient will have a maximum there. Above P_C, the activation energy will be the energy required to make a four-fold coordination in a six-fold coordinated system, which is a decreasing function of pressure, so that the diffusion coefficient now decreases with pressure. Although in real situations the average coordination number should be somewhere between four and six, the present picture will still hold.

4 MD Simulation of Silica Melt

The conjecture in the last section may be confirmed if we can demonstrate a correlation between the diffusion coefficient and the activation energy for the coordination change. The interatomic potential we have developed

Table 1: Total energy of each polymorph at 300K for three types of potential parameters.

Type	Energy (kcal/mol)			Energy difference (kcal/mol)
	quartz	coesite	stishovite	(stishovite)−(coesite)
I	−1231.52	−1235.15	−1226.86	8.29
II	−1249.64	−1253.97	−1238.40	15.6
III	−1270.73	−1273.97	−1250.27	23.7

has two virtues for this purpose. First, the potential can reproduce the structures and bulk moduli of both four-fold and six-fold coordinated polymorphs of silica. Second, we can tune the relative stability of four-fold and six-fold coordinated polymorphs by a slight modification of a potential parameter without losing the first virtue.

The interatomic potential has the following functional form,

$$U_{ij}(r) = Q_iQ_j/r + f_0(b_i + b_j)\exp[(a_i + a_j - r)/(b_i + b_j)] - C_{ij}/r^6,$$

where r is the distance between the ith and the jth atoms and $f_0 =$ 1kcalÅ^{-1}mol^{-1}. The potential parameters used in the simulation in Section 2 are $Q_{\mathrm{Si}} = +2.4$ and $Q_{\mathrm{O}} = -1.2$ in elementary charge, $a_{\mathrm{Si}} = 0.86$, $a_{\mathrm{O}} = 2.05$, $b_{\mathrm{Si}} = 0.176$ and $b_{\mathrm{O}} = 0.033$ in Å, $C_{\mathrm{SiSi}} = 0$, $C_{\mathrm{SiO}} = 1633$ and

$C_{OO} = 4956$ in kcalÅ^{6}mol^{-1}. As the relative stability of four-fold and six-fold coordinated systems is sensitive to the atomic radius ratio a_{Si}/a_O, we purposely vary this ratio to tune the relative stability to investigate its correlation with the diffusion coefficient.

We consider three sets of potential parameters, where $a_{Si} = 0.87$ Å(type I), $a_{Si} = 0.86$ Å(type II) and $a_{Si} = 0.85$ Å(type III), and the values of other parameters are fixed as above. The total energies of low-quartz, coesite and stishovite at room temperature and ambient pressure are summarized in Table 1 together with the energy difference of the most stable four-fold system (coesite in this case) and the six-fold system (stishovite).

The critical pressure P_C defined in the last section is roughly estimated as a common tangent of the free energies against volume for coesite and stishovite. To do so, we first approximate the free energy $E - TS + pV$ by the enthalpy $E + pV$, neglecting the entropy differences. Here E is the internal energy tabulated in Table 1. The curvatures of the free energies against V around the minimum are given by the bulk moduli of coesite and stishovite: we have employed a common bulk moduli (100 GPa for coesite and 300 GPa for stishovite) for the three types of parameter sets, since they are rather insensitive to a_{Si}. The P_C's estimated in this way are 6 GPa for type I, 12 GPa for type II and $\leq$ 21 GPa for type III parameter sets, respectively. Experimentally, the transition pressure is reported to be about 8 GPa by experiment [16].

The self-diffusion coefficient, D, can be determined from the linear time-dependence of the mean square displacement of atoms of each species in the MD simulation using Einstein's relation,

$$\langle (\vec{r}(t) - \vec{r}(0))^2 \rangle = 6Dt + const.,$$

where $\vec{r}(t)$ denotes the position of one particle at time t and $\langle \cdots \rangle$ means the average over particles of the species considered.

Starting from crystalline (low-cristobalite) configuration for the ambient pressure we melt the system at 4000 K for 100 psec in the MD run. The pressure is then increased by 4 GPa up to 36GPa. The pressure was controlled by keeping the cubic symmetry of the basic cell and the temperature was controlled by a simple scaling of momenta in this simulation. After at least 20 psec (10000 MD steps) equilibration at each pressure, the diffusion coefficient is determined from the data accumulated over 60 psec. The above process is repeated for each of the three parameter sets. The number of particles contained in the basic cell for the periodic boundary condition is 324. We have also looked into a larger system with 576 particles to check the size effect and found almost no quantitative difference at least at low pressures. The only difference is

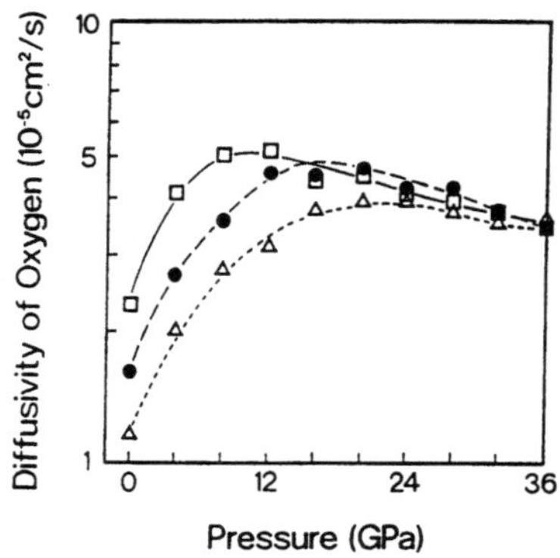

Figure 4: Pressure dependence of the self-diffusion coefficient of oxygen atoms at 4000 K. Three kinds of symbols and lines correspond to three types of potential parameters: type I (□,———), type II (●,– – –) and type III (△,-------). The lines are guides to the eye.

that the fluctuation in D against time is relatively small for the larger system size.

The key result for the pressure dependences of D for oxygen at 4000 K for the three types of parameter sets are shown in Fig.4. Though the results for silicon are not shown here, they are qualitatively very similar to those for oxygen. A common characteristic for the three parameter sets is that the diffusion coefficients show maxima as a function of pressure. This is exactly the anomalous diffusion property. Moreover, there exists a conspicuous correlation between the pressures which give the highest diffusivity and the P_C's estimated by the above discussion. This serves as clear evidence that the mechanism for the atomic diffusion in silica melt presented in the last section is indeed in action. The details of the simulation will be published elsewhere [17].

5 Conclusion

We have provided and tested a conjecture on the diffusion process in the melts of silica and framework silicates: the atomic diffusion arises from the transient increase of the coordination number of silicon atoms, which results in rearrangement of the framework structure. The anomalous diffusion property is explained in a natural way in this picture.

The main message of the present study is that the MD method with non-empirical interatomic potentials enables us to give qualitative discussions on the structure of condensed phases, and to look further into mechanisms of dynamical processes by controlling the potential.

Acknowledgment
The author wishes to acknowledge collaboration and discussion with Y. Matsui, H. Aoki and M. Tsukada. He is also indebted to S. Nosé and M. Matsui for original MD programs and I. Kushiro and N. Shimizu for useful discussion. Numerical calculations were performed at the Computer Centre, the University of Tokyo. This work has been supported by a Grant-in-Aid from the Ministry of Education, Science and Culture, Japan.

References

[1] I. Kushiro, *J. Geophys. Res.* **73**, 619-634(1976).

[2] I. Kushiro, in *Physics of Magmatic Processes* ed. R. B. Hargraves (Princeton University Press, 1980), pp.93-120.

[3] N. Shimizu and I. Kushiro, *Geochim. Cosmochim. Acta* **48**, 1295-1303(1984).

[4] C. M. Scarfe, B. O. Mysen and D. Virgo, *Carnegie Inst. Wash. Year Book* **78**, 547(1979).

[5] C. A. Angell, P. A. Cheeseman and S. Tamaddon, *Science* **218**, 885-887(1982): *Bull. Minéral.* **106**, 87-97(1983).

[6] C. A. Angell, P. A. Cheeseman and R. R. Kadyyala, *Chemical Geology* **62**, 83-92(1987).

[7] I. Kushiro, *Carnegie Inst. Wash. Year Book* **77**, 672-674(1978).

[8] S. Tsuneyuki, M. Tsukada, H. Aoki and Y. Matsui, *Phys. Rev. Lett.* **61**, 869-872(1988).

[9] S. Tsuneyuki, H. Aoki, M. Tsukada and Y. Matsui, *Phys. Rev. Lett.* **64**, 776-779(1990).

[10] S. Tsuneyuki, Y. Matsui, H. Aoki and M. Tsukada, *Nature* **339**, 209(1989): S. Tsuneyuki, H. Aoki and Y. Matsui, *Molecular Simulation*, in press.

[11] R.J. Hemley, A.P. Jephcoat, H.K. Mao, L.C. Ming and M.H. Manghnani, *Nature* **334**, 52-54(1988).

[12] M. Parrinello and A. Rahman, *J. Appl. Phys.* **52**, 7182-7190(1980).

[13] S. Nosé, *J. Chem. Phys.* **81**, 511-519(1984).

[14] A very recent experimental result (Y. Tsuchida and T. Yagi, *Nature* **347**, 267-269(1990)) shows that low-cristobalite indeed undergoes pressure-induced structural transformation, twice, at pressures

of about 10 GPa and 30 to 40 GPa, though the newly obtained structures do not coincide with those predicted by the MD simulation.

[15] E.M. Stolper and T.J. Ahrens, *Geophys. Res. Lett.* **14**, 1231-1233(1987).

[16] T. Yagi and S. Akimoto, *Tectonophys.* **35**, 259-270(1976).

[17] S. Tsuneyuki, H. Aoki and Y. Matsui. in preparation.

Interatomic Potential Models for Molecular Dynamics Simulations of Multi-component Oxides

K. Kawamura

Department of Chemistry, Faculty of Science,
Hokkaido University, Sapporo 060, Japan

Abstract. Two sets of interatomic potential models for multi-component oxide systems were proposed for the use of molecular dynamics (MD) simulations under normal pressure. The models involve Coulomb and repulsion terms (Model I) and an additional Morse term (Model II). The parameters of O, Na, Mg, Al, Si, K, Ca, Ti, Sr and Ba ions for Model I and B, O and Si for Model II were determined by trial and error procedures using MD simulations of various oxide crystal structures. The criteria of the determinations are reproducibility of the crystal structures and the calculated pressure in constant cell MDs. Most of the structures were reproduced within mean square displacements of 0.02Å^2 from the experimentally determined atomic positions. The models, therefore, are considered to be suitable for structural studies of multi-component oxide crystals, liquids and glasses containing the above ions under normal pressure. Some applications using the models are shown.

1. Introduction

Multi-component silicates appear in the earth sciences as crystals, liquids and glasses. They are major components of the earth's crust and mantle. Inorganic material sciences are often required to treat complex oxide compounds for various kinds of new devices such as high temperature superconductors, non-linear optical devices, etc. Therefore, it is desirable to establish methods which treat multi-component oxide systems. Molecular dynamics (MD) is one of the most effective tools for systems of large numbers of atoms. To make MD effective, construction of practical interatomic potential models for multi-component oxides is essential.

Woodcock et al.[1] and Angell et al.[2] developed and used a set of potential parameters (WAC model) of O, Na, Al and O for high temperature silica and silicate melts (Na_2O-SiO_2, Na_2O-Al_2O_3-SiO_2). This model was considered not to be effective for the crystals because the parameter determinations were made using glass or melt structures (radial distribution functions).

In this study, the parameters in interionic potential functions are estimated and refined using MD calculations of various crystal structures, and tested for some multi-component oxide systems.

2. Procedures of molecular dynamics calculations

A rectangular parallelepiped basic cell and three-dimensional boundary conditions were adopted for the present MD calculations for crystalline systems in which the crystal systems were restricted to those other than monoclinic and triclinic. Hexagonal lattices were converted to rectangular ones. The Ewald method was applied for the calculations of electrostatic potential and force. Integration of motions of ions was carried out using the Verlet algorithm ($\Delta t=2.5\times10^{-15}$ s). Temperature control was performed by particle velocity scaling. [3]

A simplified procedure for pressure controlled MD calculations is involved in MD programs [4,5]. The following procedures are applied for MD systems having a rectangular basic cell. Pressure and the diagonal components of the tensor are used to adjust the cell volume and shape to a prescribed pressure. Using the procedure given above, extra computations are rarely required and pressure is controlled reasonably well. The correct functioning of the program was confirmed by obtaining the same results as from constant pressure, constant temperature MD calculations [6,7,8].

Table 1. Systems of multi-component oxide crystals for MD calculations. *'s as superscripts to the crystal structure names denote high pressure phases. N_a, N_b and N_c are the number of stacked unit cells to make a basic cell. N's are the number of ions in a basic cell. Py-garnet means pyrope ($Mg_3Al_2Si_3O_{12}$), Gr-garnet grossularite ($Ca_3Al_2Si_3O_{12}$)

Crystal structures	Unit cell parameters a/Å	b/Å	c/Å	N_a	N_b	N_c	N
SiO_2:low-quartz	4.913	8.510	5.405	3	2	3	324
				4	3	4	864
				5	3	5	1350
SiO_2:low-cristobalite	4.979	4.979	6.950	3	3	2	216
				4	4	3	576
				5	5	4	1200
SiO_2:stishovite*	4.177	4.177	2.665	4	4	6	576
α-Mg_2SiO_4:forsterite	4.753	10.190	5.978	3	2	3	504
β-Mg_2SiO_4*	5.698	11.438	8.257	3	2	2	672
γ-Mg_2SiO_4*	8.065	8.065	8.065	2	2	2	448
$MgSiO_3$:enstatite	18.230	8.817	5.181	1	2	3	480
Na_2SiO_3	10.43	6.02	4.81	2	3	3	432
$K_2SiSi_3O_9$*	6.612	11.453	9.510	3	2	2	720
α-Ca_2SiO_4	5.091	11.371	6.732	3	2	3	504
α-Al_2O_3	4.758	8.241	12.991	3	2	2	720
Al_2SiO_5:sillimanite	7.488	7.681	5.777	2	2	3	384
$MgAl_2O_4$:spinel	8.081	8.081	8.081	2	2	2	448
Py-garnet*	11.456	11.456	11.456	2	2	2	1280
Gr-garnet	11.846	11.846	11.846	2	2	2	1280
B_2O_3-I	4.336	7.510	8.340	4	2	2	480
				4	3	2	720
$Mg_3(BO_3)_2$	5.398	8.416	4.497	3	2	4	528
$CaTiO_3$:perovskite	5.383	5.446	7.645	3	3	2	360
$SrTiO_3$:perovskite	3.905	3.905	3.905	4	4	4	320
$BaTiO_3$:perovskite	3.995	3.995	4.034	4	4	4	320

3. Interatomic potential model and parameter determination

Parameter sets of interatomic potential functions for the full ionic model (Model I) and partially ionic model (Model II) were estimated and refined by means of trial and error procedures using MD calculations of various crystal structures. Systems of

Table 2. The parameter set of Model I for the full ionic interatomic potential functions.

Ion	z	a/Å	b/Å	Ion	z	a/Å	b/Å
B	3	0.720	0.080	K	1	1.595	0.080
O	-2	1.629	0.085	Ca	2	1.429	0.080
Na	1	1.260	0.080	Ti	4	1.235	0.080
Mg	2	1.161	0.080	Sr	2	1.610	0.080
Al	3	1.056	0.080	Ba	2	1.819	0.080
Si	4	1.012	0.080				

Table 3. Results of constant cell MD calculations of crystal structures by Model I. P_x, etc. are diagonal components of pressure tensor

	N	P [GPa]	P tensor [GPa] P_x	P_y	P_z	Mean square displacement [Å^2] O	Si	Mg(Na,K,Ca)	
SiO_2:	324	21(2)	22	22	21	0.017	0.014		
low-qz	864	21(3)	21	21	21	0.018	0.015		
SiO_2:low-	216	18(2)	20	19	17	0.022	0.020		
cristo.	576	18(.8)	18	18	19	0.286	0.128		
(WAC)	216	16(2)	16	16	17	0.039	0.010		
SiO_2:St*	576	68(2)	67	67	77	0.003	0.002		
α-Mg_2SiO_4	504	-2(4)	2	-3	-3	0.009	0.010	0.005	
β-Mg_2SiO_4	672	-3(2)	0	0	-7	0.005	0.003	0.007	
γ-Mg_2SiO_4	448	-4(4)	-4	-4	-4	0.004	0.003	0.004	
$MgSiO_3$	480	9(2)	9	6	12	0.013	0.011	0.025	
Na_2SiO_3	432	8(3)	11	7	6	0.016	0.008	0.025(Na)	
(WAC)	432	4(2)	3	3	7	0.065	0.025	0.131(Na)	
$K_2SiSi_3O_9$	720	13(.7)	13	13	13	0.085	0.040	0.055(K)	
α-Ca_2SiO_4	504	1(2)	4	1	0	0.012	0.007	0.007(Ca)	
						O	Si	Al	
α-Al_2O_3	720	-15(.4)	-17	-17	-13	0.003		0.003	
Al_2SiO_5	384	-3(2)	-3	-10	4	0.008	0.006	0.007	
$MgAl_2O_4$	448	0(.4)	0	0	0	0.005		0.003	0.004(Mg)
Py-garnet	1280	8(.8)	8	8	8	0.007	0.003	0.003	0.016(Mg)
Gr-garnet	1280	1(.2)	1	1	1	0.005	0.003	0.003	0.004(Ca)
						O	B/Ti		
B_2O_3	480	12(1)	21	21	-6	0.019	0.039		
$Mg_3(BO_3)_2$	528	4(1)	-5	11	8	0.023	0.019	0.008(Mg)	
$CaTiO_3$	360	0(.3)	-1	2	0	0.005	0.002	0.006(Ca)	
$SrTiO_3$	360	0(.1)	0	0	0	0.004	0.002	0.006(Sr)	
$BaTiO_3$	360	0(.3)	2	2	-3	0.008	0.003	0.002(Ba)	

crystal structures used in parameter determinations are shown in Table 1. The temperature is 300 K for all systems, and pressure controlled MD calculations were carried out at 0.0001 GPa except for low-quartz at 6.8GPa.

Table 4. Results of pressure controlled MD calculations of crystal structures by Model I at 0.1 MPa.

	N	Unit cell param. [Å]			M.s.d. [Å^2]		
		a	b	c	O	Si	Mg
SiO_2:low-quartz	324	5.164	8.945	5.704	0.197	0.031	
	864	5.152	8.924	5.726	0.211	0.035	
SiO_2:low-crist.	216	5.311	5.308	7.508	0.471	0.143	
	576	5.299	5.298	7.498	0.534	0.170	
(WAC)	216	5.225	5.217	7.389	0.485	0.154	
SiO_2:stis.	576	4.235	4.235	2.727	0.004	0.002	
α-Mg_2SiO_4	504	4.787	10.096	5.940	0.009	0.011	0.006
β-Mg_2SiO_4	672	5.722	11.471	8.121	0.006	0.005	0.007
γ-Mg_2SiO_4	448	8.047	8.047	8.046	0.004	0.003	0.004
$MgSiO_3$	480	18.279	9.081	5.354	0.060	0.023	0.097
Na_2SiO_3	432	10.745	6.112	4.892	0.028	0.014	0.048(Na)
$K_2SiSi_3O_9$	720	6.791	11.757	9.669	0.024	0.007	0.046(K)
α-Ca_2SiO_4	504	5.156	11.362	6.683	0.013	0.008	0.008(Ca)

	N	a	b	c	O	Si	Al	Mg/Ca
α-Al_2O_3	720	4.736	8.202	12.975	0.003		0.003	
Al_2SiO_5	384	7.496	7.506	5.852	0.007	0.005	0.010	
$MgAl_2O_4$	448	8.082	8.082	8.082	0.005		0.003	0.004
Py-garnet	1280	11.521	11.521	11.518	0.008	0.003	0.003	0.022
Gr-garnet	1280	11.857	11.858	11.857	0.005	0.003	0.003	0.004

	N	a	b	c	O	B/Ti	
B_2O_3	480	4.606	7.975	8.122	0.057	0.020	
$Mg_3(BO_3)_2$	528	5.255	8.603	4.555	0.021	0.016	0.011(Mg)
$CaTiO_3$	360	5.368	5.461	7.643	0.005	0.002	0.007(Ca)
$SrTiO_3$	360	3.905	3.905	3.905	0.004	0.002	0.007(Sr)
$BaTiO_3$	360	4.007	4.007	4.008	0.008	0.004	0.002(Ba)

Table 5. Comparison of internal energy, density and angles of silica and Mg_2SiO_4 polymorphs calculated using Model I.

		Internal energy / kJ·mol^{-1}	Density / g·cm^{-3}	Si-O-Si angle
SiO_2	cristobalite	-13751.2(3)	1.8964(7)	178.8(2.7)°
	quartz	-13715.2(2)	2.2738(8)	160.6(3.3)°
	stishovite	-13565.0(2)	4.0813(4)	-
Mg_2SiO_4	forsterite	-22181.3(4)	3.2559(8)	
	β-phase	-22285.6(5)	3.5069(8)	
	γ-phase	-22335.5(4)	3.5874(5)	

Low cristobalite

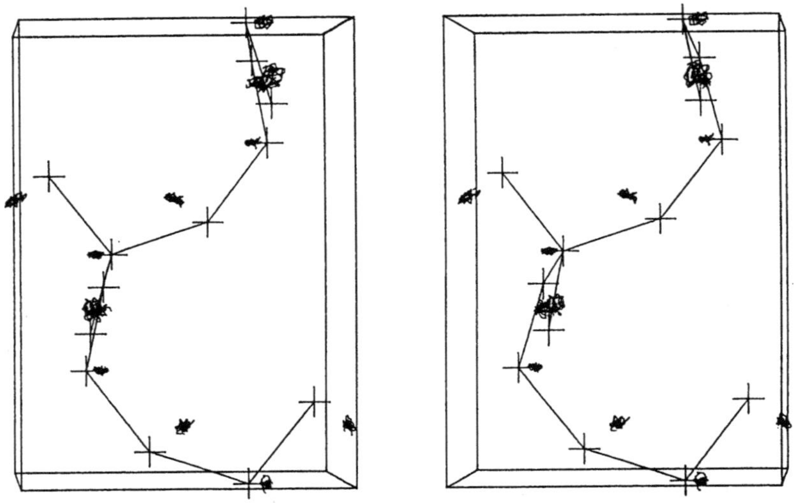

α-Mg_2SiO_4

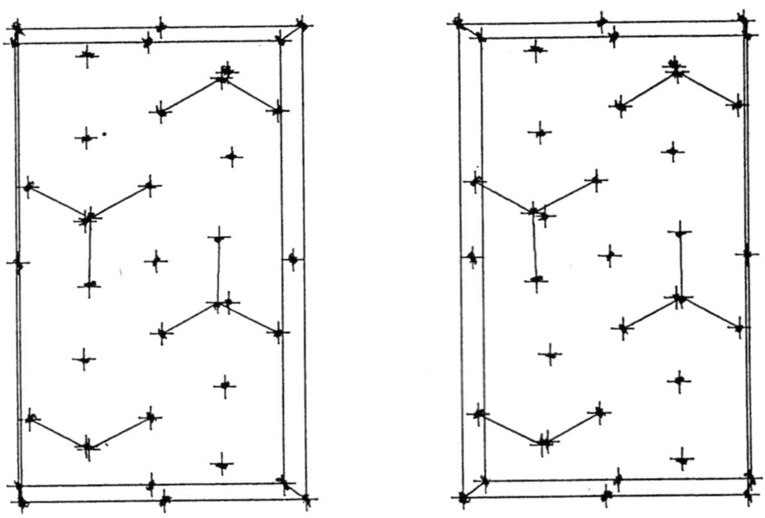

Figure 1. Stereoscopic paired view of atomic trajectories in constant cell MD simulations of low cristobalite (N=576) and α-Mg_2SiO_4 (forsterite, N=504) at 300 K with Model I (full ionic model). The unit cell of the crystal is displayed. Crosses represent the atomic positions determined by experimental structure analyses, and the length is 0.5 Å. Lines represent Si-O bonds.

3.1 Full ionic model (Model I)

Interatomic potential functions used in Model I are two-body central and composed of full ionic Coulomb and short range repulsion terms [9,10,11]:

$$u_{ij}(r_{ij}) = z_i z_j e^2/r_{ij} + f_o(b_i+b_j)\exp[(a_i+a_j-r_{ij})/(b_i+b_j)].$$

Here, $f_o = 6.948\times10^{-8}$ dyn. In this model ionic charges (z's) are full charge numbers such as z(O)=-2.

Ions and the parameters are shown in Table 2 for Model I. 22 MD systems of 20 crystal structures were used to obtain the parameters. Mean squared displacements (m.s.d.) of atoms from experimentally determined positions are within 0.04 Å^2 for most structures, whilst pressure differences between calculated and experimental (normal pressure) were within 10 GPa in most structures (Table 3). It was difficult to produce pressure close to 0.0001 GPa for all the structures using this model. The system of low cristobalite with 576 ion particles in a basic cell failed to reproduce the structure properly.

Pressure controlled MD calculations were carried out for the same systems (Table 4). The m.s.d.'s are essentially the same for most systems, except those of quartz and cristobalite, where the proper structure was no longer maintained. The comparison of internal energies for silica and Mg_2SiO_4 crystals is shown in Table 5.

The stereoscopic views of atomic trajectories in cristobalite (N=576) and in α-Mg_2SiO_4 are displayed in Fig.1.

3.2 Partially ionic model (Model II)

To improve the pressure reproducibility of Model I, adoption of partial charges of ions was found to be effective. By comparison with Model I, Model II uses partially ionic charges and an additional Morse potential term:

$$u_{ij}(r_{ij}) = z_i z_j e^2/r_{ij} + f_o(b_i+b_j)\exp[(a_i+a_j-r_{ij})/(b_i+b_j)] + f_o D_{ij}\{\exp[-2\beta_{ij}(r_{ij}-r^*_{ij})]-2\exp[-\beta_{ij}(r_{ij}-r^*_{ij})]\}.$$

Ionic charges are chosen as 60% of full charges. The Morse terms are applied only to cation-anion pairs.

Parameters of B, O and Si were determined (Table 6). The m.s.d.'s are reasonable and the calculated pressures are much better than those of Model I (Table 7(1)). It is difficult to

Table 6. The parameter set, Model II, the partially ionic interatomic potential model.

Ion	z	a/Å	b/Å	Ion pair	D/Å	β	r*/Å
O	-1.2	1.910	0.150				
Si	2.4	0.930	0.100	Si-O	81.0	2.00	1.51
B	1.8	0.620	0.100	B-O	88.0	2.00	1.25

Table 7. Results of constant cell (1) and pressure controlled (2) MD calculations of silica and boron oxide crystal structures by Model II.

(1)	N	P [PGa]	Pressure tensor / GPa P_x	P_y	P_z	M.s.d. [Å²] O	Si
quartz	324	-0.2(0.4)	0.4	0.4	-0.6	0.017	0.011
quartz (6.8GPa)	324	8.7(0.7)	8.9	8.9	8.6	0.015	0.008
cristobalite	216	0.0(0.5)	0.9	0.8	-0.9	0.042	0.029
stishovite	270	37.8(0.5)	39.8	39.7	32.7	0.005	0.004
B_2O_3-I	720	-0.1(0.4)	6.0	6.0	-12.4	0.020	0.047

(2)	N	Unit cell param. [Å] a	b	c	M.s.d. [Å²] O	Si	Internal energy [kJ·mol⁻¹]	Density [gcm⁻³]
low-	324	4.920	8.525	5.378	0.018	0.011	-5919.9(2)	2.6533
quartz	1350	4.915	8.512	5.388	0.019	0.012	-5919.4(1)	2.6561
low-	216	4.959	4.923	6.689	0.032	0.025	-5909.8(3)	2.4447
cristo.	1200	4.949	4.947	6.673	0.031	0.026	-5907.6(1)	2.4431
stisho.	270	4.326	4.326	2.674	0.004	0.007	-5837.0(2)	3.9876
B_2O_3-I	720	4.401	7.640	7.830	0.020	0.053		

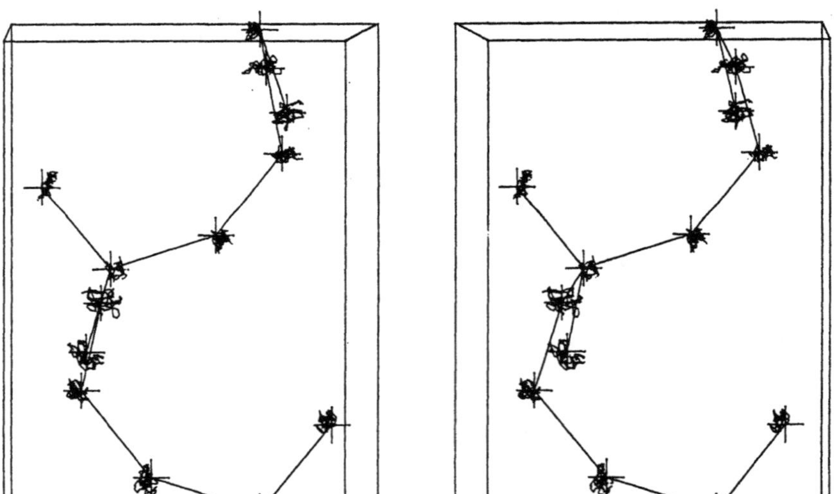

Figure 2. Stereoscopic paired view of atomic trajectories in a pressure controlled MD simulation of low cristobalite (N=1200) at 300 K and 0.1 MPa with Model II. The unit cell of the crystal is displayed. Crosses represent the atomic positions determined by experimental structure analyses, and the length is 0.5 Å. Lines represent Si-O bonds.

make the differences between components of the pressure tensor Oof B_2O_3-I crystal smaller without decreasing the structural reproducibility.

The structures are also well reproduced in pressure controlled MD calculations with large numbers of ions in a basic cell (Table 7(2)). The atomic trajectories in low cristobalite (N=1200) is shown in Fig.2

4. Application of Model I to oxide melts

Some MD calculations of silicate and borate melts using Model I are shown below.

Structures of Na_2O-B_2O_3 glasses and melts [12,13] were investigated in terms of changes in oxygen coordination numbers around boron as related to the composition. The coordination number changes from 3 fold to 4 fold partially and then from 4 fold to 3 fold with a percentage increase of Na_2O in the contents. This was determined by NMR studies of boron in the glasses. The x-ray diffraction patterns were interpreted in terms of MD-derived structures [13].

Viscosity of $Na_2O \cdot 2SiO_2$ melt [14] was studied by means of a non-equilibrium molecular dynamic method and the Green-Kubo treatment. The calculated viscosity was consistent with experimental data at higher temperatures than 1500 K.

Mixing properties of $Na_2O \cdot 2SiO_2$ - $K_2O \cdot 2SiO_2$ binary melt [15] were investigated together with the size effect of the MD basic cell. The small negative enthalpy of mixing was reproduced. The size effect of the MD basic cell was obvious. It is clear that systems of a few hundred ions in a basic cell are too small to investigate both structural and physical properties for polymerized silicate melts.

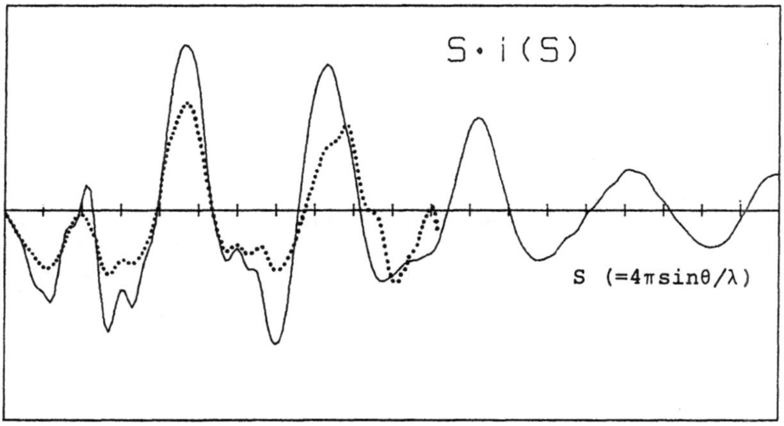

Figure 3. MD-calculated (solid line) and experimental (dotted line) [18] x-ray interference functions (Si(S)) of anorthite ($CaAl_2Si_2O_8$) melt. The vertical scale is arbitrary.

Cosmic dust with composition $MgO \cdot SiO_2$ [16] was studied. The structural and vibrational change with the density is consistent with observations and experiments [17].

A ternary oxide melt, molten anorthite ($CaAl_2Si_2O_8$) was calculated starting from a system which had 1300 ions randomly distributed in a basic cell. The calculated x-ray interference function, Si(S) was well reproduced in detail (Fig.3) [18]. The calculated structure involves many 4 membered rings composed of Si and Al tetrahedra in the network structure, and some 5 fold coordinated or octahedraly coordinated Al. The melt has a feldspar like structure rather than those of stuffed tridymite or cristobalite.

5. Discussion

The structural reproducibility of low pressure SiO_2 polymorphs using Model I is poor. The reason for this is considered to be due to excessive coulomb repulsions between Si ions tetrahedrally surrounded by oxygen atoms. This is confirmed by the large bridging (Si-O-Si) angles in the pressure controlled MD calculations compared with the experimentally determined ones (ca. 145°), as well as the small density (Table 5). The WAC potential model also showed the same behaviour (Si-O-Si=174.8° in low cristobalite with pressure controlled MD). Model II is much improved in this aspect.

The density and internal energies of Mg_2SiO_4 and silica polymorphs in pressure controlled MD calculations are listed in Table 5. The density relations are reasonable for both polymorphic series. The sequences of the internal energies, however, are difficult to accept in terms of thermochemical aspects. For example, the highest pressure phase of Mg_2SiO_4, the γ-phase, has the lowest energies. This means that Model I is difficult to use in thermodynamics studies by MD simulations.

Despite these problems, the structures of multi-component oxide melts such as Na_2O-B_2O_3 melt [12,13] and anorthite ($CaAl_2Si_2O_8$) melt are reproduced quantitatively. These show that Model I is useful for structural studies of high temperature liquids of multi-component oxides.

Using Model II, the structure reproducibility of crystals is excellent and the pressure reproducibility for quartz (normal and high pressure), cristobalite and B_2O_3-I is reasonable. However, the extension of this model to various ions is difficult and the use of this model is restricted because of the partial ionic charges.

Further developments of interatomic potential models should be continued. Models in which ionic parameters, such as charges, short range repulsion diameters, etc. are changed with their local environments may be suitable for MD calculations of multi-component oxide crystals, liquids and glasses. These parameters are difficult to determine by trial and error MD calculations. Cooperation between quantum mechanics and molecular simulations is necessary.

Acknowledgments
Part of this work was done using the mini-computer (OKITAC 50/60) of Institute for Thermal Spring Research (the present name is the Institute for Study of the Earth's Interior) of Okayama University [19]. The author is greatly indebted to Prof. Y. Matsui for giving me the opportunity to carry out this work. The author also sends his thanks to Prof. T. Yokokawa for

fruitful discussions. Part of this work was supported financially by a grant from Nihon Sheet Glass Ltd.

References

[1] L.V.Woodcock, C.A.Angell and P.Cheeseman, J.Chem.Phys. **65**, 1565 (1976).

[2] C.A.Angell, P.Cheeseman and S.Tamadon, Bull.Mineral **106**, 87 (1983). the WAC model has the following form:

$$u_{ij}(r_{ij}) = z_i z_j e^2/r_{ij} + (1+z_i/n_i+z_j/n_j)b_{ij}\exp[(\sigma_{ij}-r_{ij})/0.29],$$

where $(1+z_i/n_i+z_j/n_j)$ is the Pauling factor. Though the parameters are described for ion pairs in the paper, they seem to be pair-wise additive for ions in fact.

[3] L.V.Woodcock, in Advances in Molten Salt Chemistry (J.Braunstein, G.Mamantov and G.P.Smith, eds.), Plenum, New York (1975) p.1.

[4] K.Kawamura, PhD thesis (1985). Pressure control is performed using the following equation:

$$L_x = L_x[1+k_L(P_x-P_{set})],$$

where k_L is a variable parameter depending on pressure, temperature, history, etc. The adjustment of cell edge length is carried out like PID (proportional, integral and differential) control in temperature controls.

[5] Y.Matsui and K.Kawamura, in High Pressure Research in Mineral Physics (eds. M.H.Manghnani and Y.Syono), Terra Sci. Publ. Co., Tokyo (1987) p.305.

[6] M.Parinnello and A.Raman, Phys.Rev.Lett. **45**, 1196 (1980).

[7] S.Nose, Mol.Phys. **52**, 255 (1984).

[8] S.Nose, J.Chem.Phys. **81**, 511 (1984).

[9] Y.Ida, Phys.Earth Planet.Interiors **13**, 97 (1976).

[10] T.L.Gilbert, J.Chem.Phys. **49**, 2640 (1968).

[11] W.R.Busing, Oak Ridge National Laboratory, ONRL-5747 (1981).

[12] Q.Xu, K.Kawamura and T.Yokokawa, J.Non-Cryst.Solids **104**, 261 (1988).

[13] Q.Xu, K.Kawamura and T.Yokokawa, Nippon Yogyo-Kyokai-Shi **97**, 1416 (1989).

[14] H.Ogawa, Y.Shiraishi, K.Kawamura and T.Yokokawa, J.Non-Cryst.Solids **119**, 151 (1989).

[15] K.Kawamura, Mol.Simul. in press.

[16] A.Tsuchiyama, and K.Kawamura, IAU Proceedings (Kyoto) in press.

[17] C.Koike and A.Tsuchiyama, IAU Proceedings (Kyoto) in press.

[18] M.Okuno and F.Marumo, Miner.Journ. **11**, 180 (1982).

[19] Most of the MD calculations in this study were carried out using mainly 32 bit personal computers. Some of the large systems with more than 1000 ions in a basic cell were calculated using a RISC work station.

Transport and Dynamical Correlations in Glassy States and the Liquid–Glass Transition of Li_2SiO_3

J. Habasaki[1], *I. Okada*[1], *and Y. Hiwatari*[2]

[1]Department of Electronic Chemistry, Tokyo Institute of Technology at Nagatsuta, Nagatsuta 4259, Yokohama 227, Japan
[2]Department of Physics, Faculty of Science, Kanazawa University, Kanazawa 920, Japan

Abstract. Molecular Dynamics (MD) simulation of the Li_2SiO_3 system has been performed to learn the relationship between structure and dynamical properties near the liquid-glass transition. Some aspects of dynamics in a glassy state are shown by the self-part of the van Hove functions and by the Fourier transform of the velocity autocorrelation functions for each atomic species. A geometric structural analysis with polyhedra made of O atoms around Li^+ ions was applied to characterize the structure and dynamical behavior above and below the glass transition temperature. The Li^+ ions are trapped in the closed packed polyhedra whose shapes do not frequently change.

1. Introduction

Molecular dynamics simulations have been extensively performed to study the structures and dynamics of glasses. Some characteristic feaures of the structures have been found by using the Voronoi polyhedron analysis or Delaunay tetrahedron analysis, although there is no clear difference in types of polyhedra between liquids and glasses [1,2]. Some diagnostics of the structural slowing-down near the liquid-glass transition have also been studied [3,4]. However, the relation with the characteristic structure of the glass has been little studied. In a previous study of Li_2SiO_3[5], the number of vertices, V, of a polyhedron within the distance R_L from a Li^+ was analyzed in relation with the number of the O-O pairs, N, shorter than R_O; R_L is the position where the first plateau of the running coordination number $n_{Li-O}(r)$ ends, and R_O is the position where the second plateau of the running coordination number $n_{O-O}(r)$ ends. This analysis was found to be useful to clarify the geometrical characteristics and dynamics of the system.

The system Li_2SiO_3 forms one of the typical alkali silicate glasses, which consists of chain structures made by SiO_4 units and mobile Li^+ surrounded by O atoms. Since the structure in a neighboring range has a very short relaxation time, these local structures are considered to reach a local equilibrium very quickly even below the glass transition temperature, while the whole system is not in equilibrium.

2. Method

A MD calculation was performed in a manner similar to that in the previous study [5]. In a periodic cube of the simulation, 144 Li, 72 Si and 216 O particles were disposed. A Gilbert-Ida-type potential function [6] plus an r^{-6} term was used:

$$\Phi_{ij}=z_i z_j e^2/r+f_0(b_i+b_j)\exp\{(a_i+a_j-r)/(b_i+b_j)\}$$
$$-c_i c_j/r^6, \qquad (1)$$

where z is the effective charge number, *e* the elementary charge; a, b and c are the parameters characteristic of the atoms; f_0 is a normalization constant (4.184 kJ $Å^{-1}mol^{-1}$). The potential parameters, which are given in Table 1, were chosen to reproduce well the interference functions, si(s), obtained by X-ray diffraction [7]. The system was equilibrated at 4000 K during a run of more than 10000 time steps starting from a random configuration and the system was cooled down to obtain lower temperature samples. Simulations were carried out at 3000, 2000, 1673, 1173, 973, 700 and 500 K. The time step was 1 fs from 4000 to 2000 K and 4 fs from 1673 K to 500 K. The velocities of all particles were set to zero at the beginning of each temperature, and 200 steps were computed at constant temperature. In this note the results are given for 1673

Table 1. Parameters of the potentials used in this work

Ion	z	$c/Å^3kJ^{1/2}mol^{-1/2}$	a/Å	b/Å
O	-1.40	54.0	1.81	0.142
Si	2.60	0	0.75	0.036
Li	0.80	0	0.85	0.040

K(melt) and 700 K(glass) with the simulation data over 6000 steps and 60000 steps, respectively, under constant energy conditions. The volume at each temperature was taken from data [8,9] on the density of the melt and the glass.

3. Results and Discussion

3.1 Dynamical and Transport Correlations

The order of the diffusion coefficients in this system is $D_{Li} >> D_O \simeq D_{Si}$ both in the molten and glassy states. The self-parts of the van Hove functions defined by Eq. (2) are calculated to characterize the single particle motion in glass:

$$G_s^{\alpha}(r,t)=(1/N_{\alpha}) \sum_{i=1}^{N_{\alpha}} \langle \delta(r_i(t)-r_i(0)-r) \rangle, \quad (2)$$

where $r_i(t)$ denotes the position of species α at time t.

The function decays differently above and below the transition temperature. The functions $4\pi r^2 G_s(r,t)$ at 700 K are shown in Figs. 1 (a) for Li^+ and (b) for O. The position of the main peak for Li^+ is independent of time in the glassy state, while its area decreases gradually with the elapse of time. The tail appears at larger r and gradually increases with time. The dynamical behavior of this single particle motion of Li^+ is similar to that of a simple binary alloy system reported by Roux *et al.*[3], where the atoms diffuse *via* activated processes like individual or correlated jumps [10].

On the other hand, for O atoms, the main peak is nearly frozen and no tail appears at a longer r region at least within 192 ps. Thus, the chain structures made of SiO_4 units are little changed during this period in the glassy state, and Li^+ moves among equilibrium sites surrounded by the O atoms.

The power spectrum $fs(\omega)$ is obtained by Fourier transform of the velocity autocorrelation function. The results for 1673 and 700 K are shown in Fig. 2. The motion of Li^+ is found to correlate strongly with those of O and Si at both temperatures. The peak for Si and O at about 180-190 ps^{-1} ($\sim$1000 cm^{-1}) is attributed to the Si-O stretching mode [11].

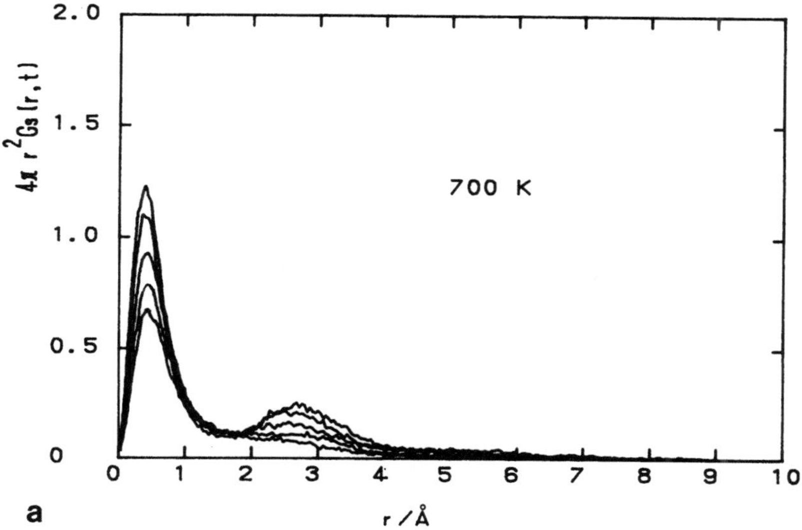

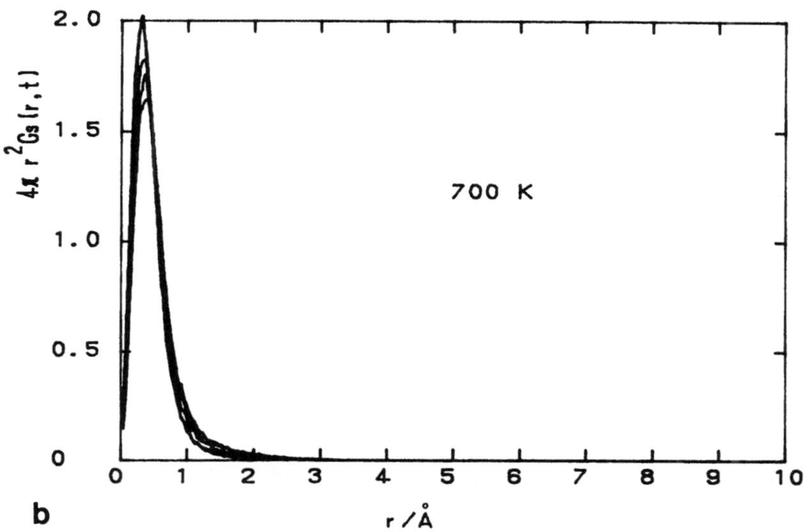

Fig. 1
(a) $4\pi r^2 G_s^{Li}(r,t)$ versus r at 700K. The curves from top to bottom for the first peak correspond to t=32, 64, 96, 128 and 192 ps, respectively.
(b) $4\pi r^2 G_s^{O}(r,t)$ versus r at 700K. The curves from top to bottom correspond to t=32, 64, 96, 128 and 192 ps, respectively.

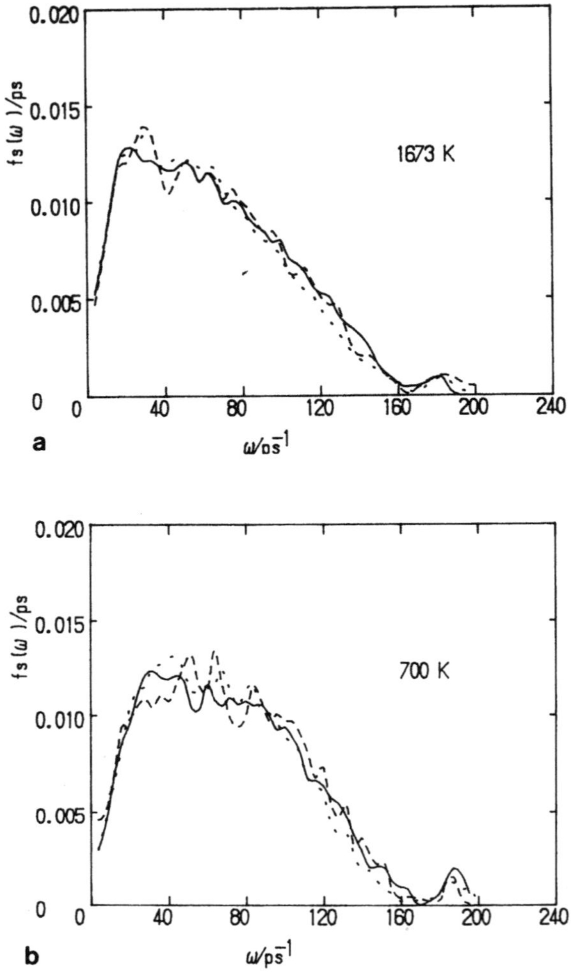

Fig. 2
Power spectra of Li^+ (full curve), Si (broken curve) and O (dotted curve) (a) at 1673 K and (b) at 700 K.

3.2 Geometric Characteristics of Structures and Motion of Lithium Relative to the Surrounding Oxygen Atoms in Glass

In the previous work [5], it has been shown that the number of vertices, V and the number of contact pairs of each polyhedron, N, are useful parameters to characterize the geometric structure of the system. Some typical structures with different V and N values are shown in Fig 3.

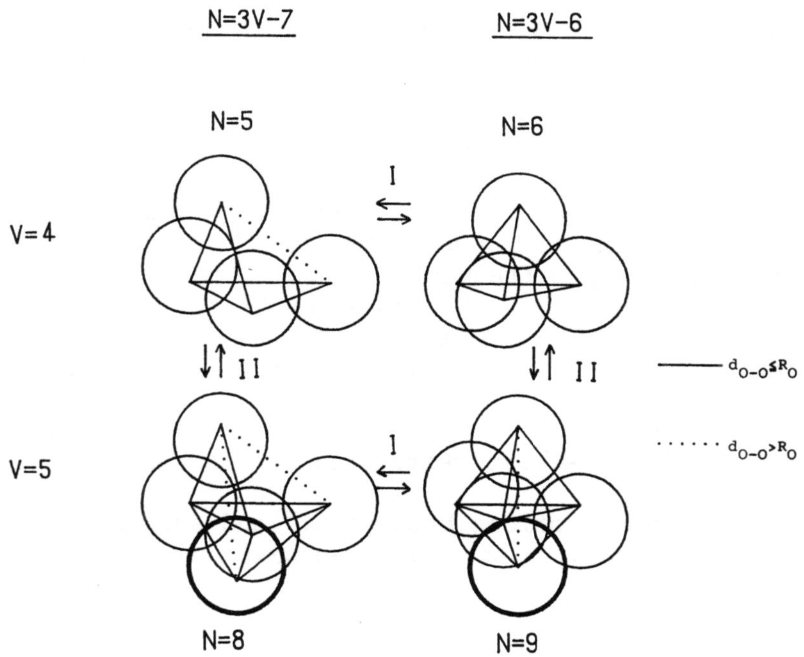

Fig. 3
Schematic drawing of some structures of the polyhedra with several values of N and V. The arrows in the figure show the main routes of structural changes in a short time interval.
The circles represent O atoms. The bold ones refer to O atoms entering or leaving the polyhedra considered.

The N values were found to have maxima at N=3V-6 in the glassy state. This means that the O atoms are packed tightly. On the other hand, the structures N=3V-7, which have free space between two O atoms, were observed frequently in the liquid state. Each polyhedron can be easily deformed until the condition N=3V-6 is achieved. When N equals 3V-6, the polyhedron has no more O-O pairs which can be shortened independently of other O-O pairs in the same polyhedron.

The O atoms consisting of a polyhedron can generally move with cooperative motion or with changing the shape of the polyhedron. Since a polyhedron shares the O atoms with other polyhedra or other SiO_4 units, it is hard to move with cooperative motion with other polyhedra at low temperature due to limited free space. The change in shape of the polyhedron is also restricted by the limited free space. Thus the changes in the distribution pattern of N values are related to the changes in dynamical properties of the O atoms as well as statical properties. The motion of Li^+ correlates of course with this structural change.

The relation between the structural fluctuation of the polyhedron and the motion of Li^+ has been analyzed. By computing the transition probability of N values, two types of changes were frequently found, namely

I. $\Delta N = \pm 1$, $\Delta V = 0$; II. $\Delta N = \pm 3$, $\Delta V = \pm 1$ (see the corresponding change in Fig. 3). The time interval was taken during which N was changed for half of the polyhedra, that is $\Delta t = 4$ fs at 1673 K and 16 fs at 700 K.

Also, we have found that the translational (diffusive) motions of Li^+ are almost restricted to the case that II takes place.

The relation of motions between Li^+ and surrounding O atoms are examined with the time-averaged structures over

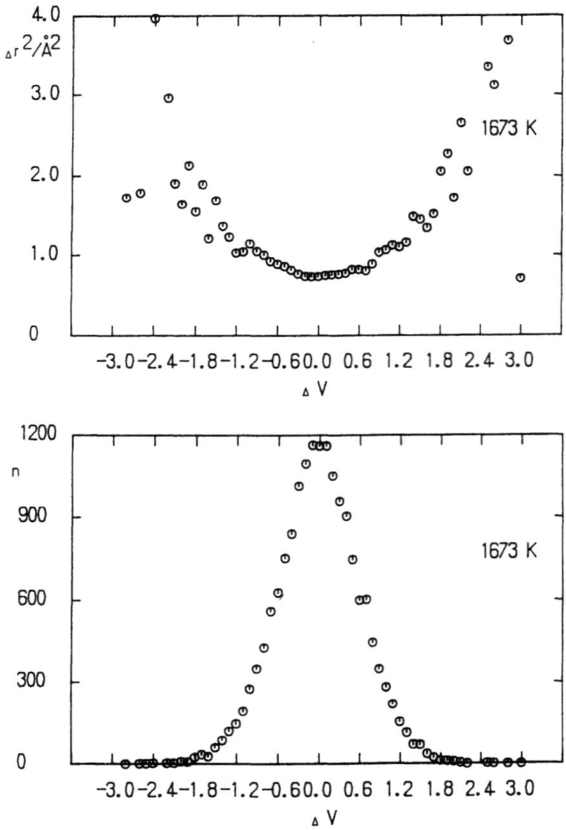

a

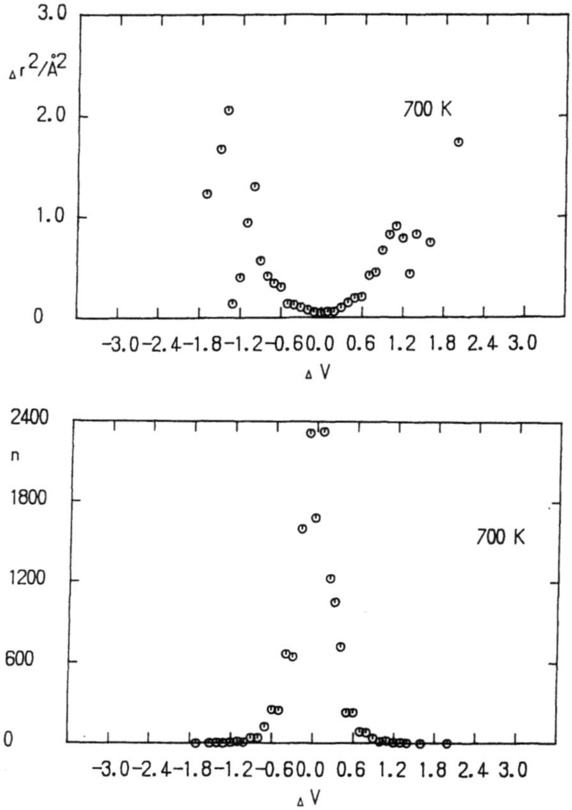

b

Fig. 4
Relation between changes in V and mean square displacements for Li^+ trapped in the polyhedron (a) at 1673 K and (b) at 700 K. Each value is calculated with the time averaged structures over 0.2 ps (1673 K) and 0.8 ps (700 K). The corresponding distribution of n is also shown.

several times of vibrational frequencies. The relation between ΔV and $\Delta r^2 = (\Sigma r_i^2/n)$ of Li^+ is shown in Fig. 4 together with distribution of n, where r_i^2 is the squared displacement of Li^+ and n is the number of Li^+ ions contained in $\Delta V \pm 0.05$. Both ΔV and r_i^2 values are calculated with the time-averaged position of atoms over 0.2 ps for 1673 K and over 0.8 ps for 700 K. It turns out from these figures that for larger $|\Delta V|$, larger mean square displacements of Li^+ are obtained. The correlation is not distinct for a large ΔV region because of poor statistics and the fact that the changes in V are not necessarily additive

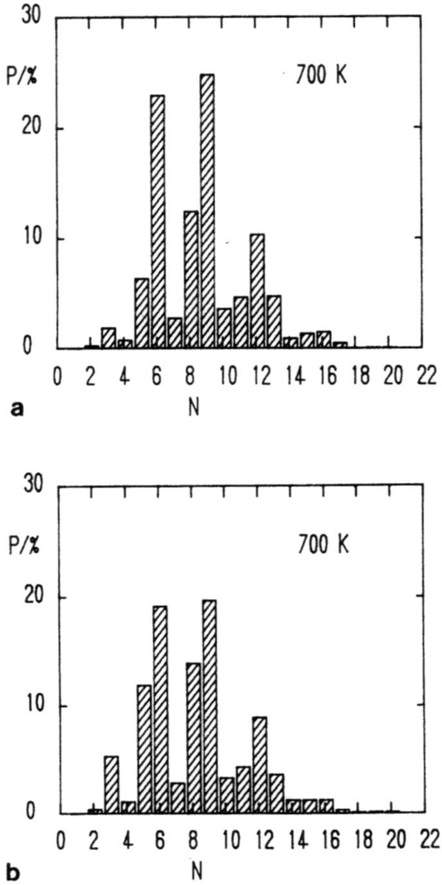

Fig. 5
The distribution of N values for the polyhedra belonging to each class, specified by the ΔV values in Fig 4 (b). (a) class 1 ($|\Delta V| \leq 0.5$) and (b) class 2 ($|\Delta V| > 0.5$). (N=2~3 for V=3, N=4~6 for V=4, N=7~10 for V=5, N=11~13 for V=6, N=14~17 for V=7).

for the motion of Li^+. The sharp distribution of n around ΔV=0 indicates the transformation of polyhedron scarcely occurs at 700 K as expected from the distribution of N which we have already mentioned, and this results in a small motion of Li^+. A similar relation between ΔV and Δr^2 was obtained at 1673 K, where the distribution of n is broader than that for the glass. The distribution of n should correlate with the diffusion of Li^+, although other factors may also affect it.

To study the relation between the changes in the shape of the polyhedra and the distribution of N values, polyhedra at 700 K were classified into two groups by their ΔV values in Fig. 4 (b). The distribution of N values of the class 1, attributed to small ΔV values ($|\Delta V| \leq 0.5$) is shown in Fig. 5 (a), while that of the class 2 with large ΔV values ($|\Delta V| > 0.5$) is shown in Fig. 5 (b). The structure with N=3V-6 is found more frequently in the class 1 than in the class 2. Thus, the geometrical structure characterized by N is related with the diffusibility of Li^+.

3.3 Conclusion

In order to characterize the dynamical slowing-down in the vicinity of the glass transition in Li_2SiO_3, changes in the geometrical structure and related dynamics were examined. Geometrical structural changes in the polyhedra of O atoms centered on each Li^+ were found to correlate well with the diffusibility of the Li^+ in both molten and glassy states. The transformation of the polyhedron scarcely occurs in the glassy state as expected from the distribution of N, which results in a small diffusive motion of Li^+.

Acknowledgements

Two of us (J. H. and Y. H.) thank professor Sidney Yip for his useful discussions during the symposium. Part of the calculations for this work were performed with HITAC M-680 and S-820 at the Institute for Molecular Science at Okazaki. The CPU time made available is gratefully acknowledged.

References

[1] M. Tanemura, Y. Hiwatari, H. Matsuda, T. Ogawa, N. Ogita and A. Ueda, Prog. Theor. Phys. **58**, 1079 (1977).

[2] Y. Hiwatari, T. Saito and A. Ueda, J. Chem. Phys., **81**, 6044 (1984).

[3] J. N. Roux, J. L. Barrat and J. P. Hansen, J. Phys.: Condens. Matter **1**, 7171 (1989).

[4] J. J. Ullo, S. Yip, Phys. Rev. Lett. **54**, 1509 (1985).

[5] J. Habasaki, Molec. Phys., **70**, 513 (1990).
[6] Y. Ida, Phys. Earth Planet. Interiors **13**, 97 (1976).
[7] Y. Waseda and H. Suito, Transactions ISIJ, **17**, 82 (1977).
[8] J. O'M. Bockris, J. W. Tomlinson and J. I. White, Trans. Faraday Soc., **52**, 299 (1956).
[9] H. Hasegawa, K. Adachi, I. Yasui and M. Imaoka, Yogyo-Kyokai-Shi, **88**, 51 (1980)(in Japanese).
[10] H. Miyagawa, Y. Hiwatari, B. Bernu and J. P. Hansen, J. Chem. Phys. **88**, 3879 (1988).
[11] S. H. Garofalini, J. Chem. Phys., **76**, 16 (1982).

Part IV

Fluids and Hydrodynamics

Simulation of Sub-molecular and Supra-molecular Fluids

D. Frenkel

F.O.M. Institute for Atomic and Molecular Physics, P.O. Box 41883,
1009 DB Amsterdam, The Netherlands

Abstract

In this lecture, I discuss two 'hard' problems in the numerical simulation of many-body systems. The first is the quantitative study of the slow (algebraic) decay of tagged-particle correlation functions in simple fluids. The second is the numerical study of phase equilibria in polymer systems.

Using conventional simulation techniques, many aspects of these problems cannot be studied with existing (super)computers. However, if we consider 'lattice-gas' versions of the same problems, then it turns out that a speed-up of 6-10 orders of magnitude can be achieved using a very simple algorithm. As a result, we can now test theoretical (mode-coupling) predictions for long-time tails with unprecedented accuracy. Similarly, certain phenomena in polymer physics, such as the polymer-induced clustering of colloids, that did thus far defy simulation, can now be observed and studied numerically.

1 Introduction

The algorithms that we use in Monte Carlo and Molecular Dynamics simulations have not changed fundamentally since they were first introduced in the fifties and early sixties [1]. This fact clearly illustrates something that hardly needs illustrating, namely that the 'founding fathers' of computer simulation were exceptional scientists. At the same time, this observation is a bit depressing. Even after 35 years of rapid growth in the number of applications of computer simulation to problems in physics and chemistry, the main increase in computing speed is still simply due to the increased speed of computers. In other words: there are no short-cuts in the simulation of atomic and molecular fluids. The number of floating-point operations in a typical simulation may be somewhat larger than strictly necessary, but not by orders of magnitude.

This is indeed depressing news, because even with today's supercomputers, there are numerous properties of atomic or (macro)-molecular systems that defy numerical simulation. In this lecture I shall focus on two examples. The first example has to do with the numerical study of the velocity autocorrelation-function of a tagged particle in a fluid. The second concerns the properties of a mixture of colloidal particles and polymer molecules. At first sight, these two problems appear totally unrelated. However, I hope to show that there is an intimate connection between them. Before doing so, let us first consider the individual problems.

1.1 Long-time tails

In the history of the kinetic theory of fluids, 1969-1970 was a crucial year. In that year Alder and Wainwright [2] published a paper in which they demonstrated the breakdown of the 'Molecular Chaos' assumption. The Molecular Chaos assumption, originally introduced by Boltzmann as the 'Stoßzahlansatz', states that the collisions experienced by a molecule in a fluid are uncorrelated. One consequence of this assumption is that the velocity autocorrelation function (VACF) of a tagged particle in fluid should decay exponentially. What Alder and Wainwright found is that the VACF of a particle in a moderately dense fluid of hard spheres or hard disks does not decay exponentially but algebraically. These algebraic long-time tails are the consequence of coupling between particle diffusion and shear modes in the fluid.

The Alder-Wainwright simulations caused a complete overhaul of the kinetic theory of dense fluids. The subsequent theoretical analyses of algebraic long-time tails were either based on an extension of kinetic theory [3] or on mode-coupling theory [4]. For a review, see ref. [5] . In the mode-coupling theory by Ernst, Hauge and van Leeuwen [4], it is assumed that the long-time tail is the consequence of coupling between particle diffusion and shear modes in the fluid. To a first approximation the leading term in the long-time tail of the velocity ACF is given by:

$$< v_x(0)v_x(t) > \approx \frac{D-1}{D}\frac{1}{\rho(4\pi(D_0+\nu_0)t)^{D/2}} \equiv \frac{d_0}{t^{D/2}}, \quad (1)$$

where ρ is the number density, D_0 the 'bare' self-diffusion constant, ν_0 the kinematic viscosity and D the dimensionality. These 'bare' transport coefficients are obtained by computing the diffusion constant and the kinematic viscosity using Boltzmann's molecular chaos assumption, i.e. ignoring any contribution of the long-time tails themselves. In a self-consistent mode-coupling theory, the transport coefficients that enter into the expression for the long-time tails are computed using information about the long-time tails themselves. I shall return to this point below.

Following this theoretical work, simulations were performed by Levesque and Ashurst [6] and, most extensively, by Erpenbeck and Wood [7, 8] with the aim to verify the validity of eqn. 1. For three-dimensional fluids these simulations are extremely expensive because very long simulations on very large systems must be performed. Even so, the system sizes studied in the simulations of Erpenbeck and Wood were such that it was essential to apply finite-size corrections to the corresponding mode-coupling theory before a meaningful comparison with the simulations could be made. Following such an approach, Erpenbeck and Wood found agreement between their simulation results for the VACF and a finite-size mode-coupling theory for a number of different densities. Nevertheless the statistical accuracy of their data was such that it was not meaningful to verify either the value of the exponent of the algebraic tail or the functional form of the density-dependent tail coefficient independently.

In the case of two-dimensional fluids there is another problem. Ever since the discovery of hydrodynamic tails, it has been realized that a consistent description of mode-coupling effects in a two-dimensional fluid would result in a long-time tail that decays faster than t^{-1}, because in $2D$ the self-diffusion constant D diverges. Moreover, the viscosity, which is proportional to the integral of the stress auto-correlation function, is also expected to diverge. Alder and collaborators resolved this difficulty by introducing time-dependent transport coefficients [9]. In this way, the hydrodynamic description could be made self-consistent. This self-consistent mode-coupling theory predicts that the decay of the VACF is slightly faster than t^{-1}, namely

$$\langle v_x(0)v_x(t)\rangle \sim \frac{1}{t\sqrt{\ln t}}. \tag{2}$$

Unfortunately, this correction to the t^{-1}-decay was too weak to be observed in Alder's simulations or, for that matter, in any subsequent simulation of hard disks.

Of course, one may argue that long-time tails themselves are almost impossible to detect experimentally and that this holds *a fortiori* for small corrections to these tails in two-dimensional fluids. However, the concepts that underly the self-consistent mode-coupling picture are used in many branches of physics. In that context, a direct test of this proto-typical self-consistent mode-coupling effect is of considerable importance.

In summary: the direct observation by 'conventional' computer simulation of corrections to the t^{-1} decay of the velocity ACF in a two-dimensional fluid is a problem that requires much more computing power than is presently available. Below, we shall see that with 'non-conventional' simulation techniques, the effect can, in fact, be observed.

1.2 Polymers

Let us next consider a second difficult problem in computer simulation, namely the direct simulation of phase equilibria involving polymers. Polymers are molecules with many internal degrees of freedom. The entropy associated with these degrees of freedom becomes increasingly important as the molecules grow more complex. In the limit of a long polymer chain one may even, to a first approximation, ignore the excluded volume effects, but never the conformational entropy.

This feature of complex molecules is, of course, well known and it poses a formidable challenge to computer simulators: it is essential that a numerical study of long chain molecules probes a representative sample of all accessible molecular conformations and positions. Much attention has therefore been paid to the development of efficient numerical schemes to sample polymer conformations. The recent review by Kremer and Binder [10] gives excellent account of Monte Carlo schemes to study lattice models of macromolecules. Most of the Monte Carlo techniques to sample chain conformations described in that review are based on some kind of *local* rule to rotate, change or break-and-remake bonds between polymer units. Although the existing techniques are quite sophisticated, the local nature of the algorithm results in slow *global* equilibration. Global equilibration may require large scale conformational changes of the molecules and, in the case of mixtures, appreciable changes in the local composition of the mixture. Such numerical problems are not limited to Monte Carlo simulations. In fact, although Molecular Dynamics simulations [11] have the advantage that they allow the molecules to perform collective motions in a natural fashion, the time scale for such diffusive motion is still very long.

Even more problematic than the simulation of equilibrium properties of polymer systems, is the computation of the chemical potential of chain molecules. The chemical potential is a quantity that is of central interest in the study of phase equilibria. In computer simulations of simple atomic or molecular systems, the chemical potential is often evaluated using the particle-insertion method first described by Widom [12]. This method exploits the relation between the excess chemical potential μ^{ex} (i.e. the chemical potential of a species minus the chemical potential of an ideal gas at the same density) and the change in the potential energy of the system ΔV_{test}, due to the insertion of a test particle at a random position in the fluid:

$$\mu^{ex} = -k_B T \ln < \exp(-\Delta V_{test}/k_B T) > . \tag{3}$$

The angular brackets in eq. 3 denote an ensemble average over all possible configurations of the molecules already present in the system. In Monte Carlo simulations this average has a very simple interpretation: it can be thought of as the probability of acceptance of a (virtual) Monte Carlo move which consist of the random addition of a test particle to the system of interest.

Whether or not eqn. 3 can be used in practice depends on the value of this insertion probability. If the latter is very low, the statistical accuracy is low, and long runs are needed to obtain a reliable estimate of μ^{ex}. This limitation of the method is particularly relevant for polymer simulations where a naive application of the particle-insertion method would result in extremely low insertion probabilities.

Clearly, the numerical study of the chemical potential of polymers is difficult. But even more difficult is the direct simulation of phase equilibria involving polymers. In this case, the number of chain molecules fluctuates, while the chemical potential (i.e. the osmotic pressure) is held constant. An example of considerable practical interest is a mixture of colloidal particles and (ideal) polymers. In this system, the polymers cause an 'entropic' attraction between the colloidal particles. Direct numerical simulation of this 'polymer-induced' clustering would be so time-consuming as to be almost prohibitive.

2 Towards the impossible

Having thus introduced two difficult, and apparently unrelated, problems in computational physics, I shall next show how progress can be made on both fronts using essentially the same technique. In order to do so, we first simplify both problems by discretizing space and, in the case of the long-time tails, time. For the polymer system this discretization is achieved by considering polymers as random walks on a lattice. This simplication is very popular in numerical work on polymers because it simplifies calculations appreciable while maintaining an essential feature of the problem, namely the very large number of possible conformations. Initially, we shall only consider non-self-avoiding polymers. However, I shall indicate to what extent the present approach can be (and has been) generalized to self-avoiding polymers. The problem of transport in an atomic fluid can also be cast in a space-time discretized form if we use Lattice Gas Cellular Automata (LGCA's) to model a simple fluid.

In LGCA's, the 'atoms' are constrained to move along the bonds joining the lattice sites. No two particles can move along the same bond in the same direction. The state of the lattice is completely specified by indicating which links are occupied and which are empty. This implies that lattice-gas particles are indistinguishable.

The time evolution of the system is governed by the following rules (see figure 1):

1. Propagation: all particles move in one time step (for convenience we choose $\Delta t = 1$) from their initial lattice position (say $\mathbf{X}$) to a new position ($\mathbf{X}' = \mathbf{X} + \mathbf{c}_\alpha$; where $\mathbf{c}_\alpha$ is the velocity of species α). The velocities $\mathbf{c}_\alpha$ are such that at the end of the propagation steps all particles are once more positioned at lattice sites.

2. Collision: the particles at all sites on the lattice undergo a collision that converves the total number of particles and the total momentum at each site. These local collision rules may, or may not, be deterministic.

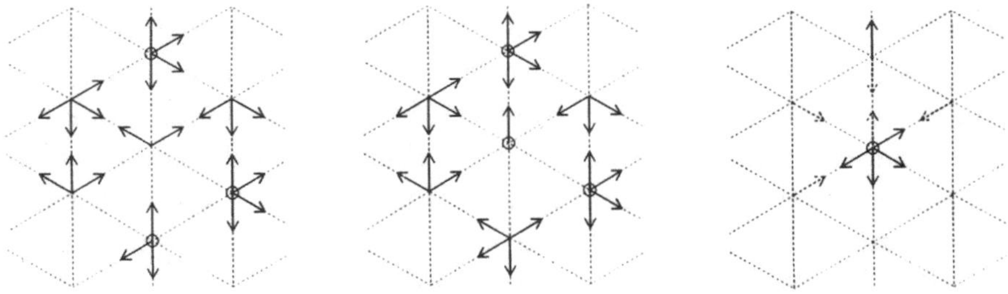

1. Illustration of the time-evolution of a two-dimensional lattice-gas cellular automaton on a triangular lattice. The model depicted here is known in the literature under the name FHP-III [13]. The arrows denote particles moving with speed 1 along the bonds linking neighboring lattice sites. The circles denote particles with speed 0. Left: typical initial configuration, just before a collision. Middle: at all lattice sites, particles collide in such a way that the number of particles and the momentum is conserved locally. Right: after the collision, the moving particles propagate to the neighboring lattice sites. This figure shows the new state of the central site. The dashed arrows show the previous positions of the particles that end up at this site.

Provided that the lattice has a sufficiently high symmetry (e.g. triangular in 2 dimensions) and the collision rules are sufficiently isotropic (for a discussion, see [14]), it can be shown that the equation that governs the time evolution of the 'flow field' of such a lattice gas becomes equivalent to the Navier-Stokes equation for an incompressible fluid in the limit that the flow velocity is much less than the particle velocity, and all spatial variations in the system occur on a scale that is large compared to the mean free path of the lattice gas particles. In this respect LGCA's model atomic fluids. In particular, we should expect to observe long-time 'hydrodynamic' tails in the velocity ACF of a tagged particle in a lattice gas.

2.1 Moment propagation

What is it that makes both the calculation of long-time tails and the simulation of polymer equilibria difficult? In the case of the long-time tails, we must sample a particular quantity (the product of the tagged-particle velocity at two different times) over many trajectories. In the polymer problem, we must count the number of allowed polymer conformations in the presence of other particles and, possibly, an external potential. However, in a conventional simulation we generate much more information than we need. For instance, in the polymer case, a Monte Carlo simulation will not only count the *number* of allowed conformations, but also their precise shape. Similarly, a Molecular Dynamics simulation will provide information about the complete history of all particles in the system, not just about one particular time-correlation function. Below, I shall indicate that it is possible to gain many orders of magnitude in computing speed, if we are willing the sacrifice such detailed, but usually irrelevant, information. This method was originally developed for the study of tagged-particle velocity-autocorrelation functions in lattice gases [15, 16, 17]. The 'moment-propagation' method allows us to compute certain averages of the n-body distribution function directly, without attempting to compute the function itself. This approach results in a very appreciable gain in computing speed: for instance, in the case of velocity autocorrelation functions in lattice gases, the method has resulted in a speed-up that varied between 10^6 and 10^{10} [15, 16]. Rather than explain how the moment propagation method works for lattice gases, I shall discuss its application to polymers, as this is simpler. To this end, let us consider a specific example of some practical interest, namely a system of *ideal*, non-interacting polymers in the external potential provided by a dispersion of 'hard-sphere' colloidal particles.

As explained above, we assume that the ideal polymer conformations correspond to random walks on a lattice. Suppose that every lattice site has b neighbors, that the polymer consists of p segments and that there are N lattice sites. Clearly, for an ideal polymer in the absence of any obstacles, the total number of allowed conformations is $\Omega_{id} \equiv N \times b^p$. If obstacles are present, some random walks are blocked, and the total number of allowed conformations, Ω_T, is less than Ω_{id}. The 'brute-force' method to determine the ratio Ω_T/Ω_{id} would be to attempt a large number of insertions of chains with arbitrary conformation at random points on the lattice. The ratio of the number of 'accepted' trial moves to the total number of attempted insertions can be used to compute the excess chemical potential of an ideal polymer chain in this system:

$$\mu_{excess} = -k_B T \log < P_{acceptance} > \tag{4}$$

This is the usual 'Widom' expression for the excess chemical potential [12]. However, unless the density of obstacles is quite low, the relative statistical error in the insertion probability will be quite large (see ref. [18]). Next, consider the 'moment-propagation' approach. To this end, let us first compute all N Boltzmann factors associated with the insertion of a point particle at any lattice site. Clearly, the sum of these Boltzmann factors is simply the partition function of a point-particle on a lattice, in an external potential. Let us denote the Boltzmann factor associated with site i as $f_i^{(0)} \equiv \exp(-u(\mathbf{r}_i)/k_B T)$. The partition function for a one-segment polymer (2 points) is computed as follows. Starting from site i, we have b ways to grow one segment. But all b directions will, in general, have different Boltzmann weights $f_j^{(0)}$, where j denotes one of the nearest neighbors of i. The total Boltzmann weight associated with the addition of a 1-segment polymer at site i is then:

$$f_i^{(1)} \equiv f_i^{(0)} \times (\sum_j^b f_j^{(0)}) \tag{5}$$

where the sum runs over the nearest neighbors of i. The partition function of a 1-segment polymer on a lattice is then simply:

$$Z_1 = \sum_{i=1}^{N} f_i^{(1)}, \tag{6}$$

(where, for convenience, I have assumed that the polymer 'head' and 'tail' are distinguishable). Repeating the same argument p times, it is clear that the Boltzmann factor associated with *all possible conformations* of a p-segment polymer starting at site i is given by:

$$f_i^{(p)} = f_i^{(0)} \times (\sum_{j}^{b} f_j^{(p-1)}). \tag{7}$$

And the total partition function is given by:

$$Z_p = \sum_{i=1}^{N} f_i^{(p)}. \tag{8}$$

The important fact to note is that the computation of Z_p, a quantity that depends on $N \times b^p$ conformations, requires only p iterations of $b \times N$ local 'propagations' of real numbers. For a fixed external potential, the partition function thus computed is *exact.* In a practical situation, as with the addition of polymer to a colloidal dispersion, the external potential is not fixed, but depends on the (continuous) coordinates of all M colloidal particles, $\{\mathbf{r}^M\}$. Hence, the total partition function of the polymer also depends on these coordinates: $Z_p = Z_p(\mathbf{r}^M)$.

The approach sketched above is not limited to completely random polymer conformations. In fact, it is simple to exclude 180° reversals or to account for the different weight of gauche and trans conformations [19]. In the latter case, instead of adding the random walks coming from all different directions to a given lattice site, we can separately store for every nearest-neighbour bond $i \to j$, the total number of walks that arrived at site i along bonds other than $j \to i$. With this bookkeeping device, it is easy to count all non-reversing random walks.

2.2 Self-avoiding polymers

In many applications, we do not only wish our polymer conformations to be non-reversing, but also self-avoiding. This latter condition cannot be implemented in a 'local' iterative algorithm based on eq. 7. Rather, the test for self-overlap has to be made after a particular non-self-avoiding polymer conformation has been selected. In practice, a large number of non-reversing random walks are sampled and the fraction of self avoiding walks (SAW's) is computed. Multiplying this fraction with the total number of non-reversing random walks, yields an estimate of the total number of SAW's.

In order to arrive at a correct sampling of the fraction of SAW's, we must generate a number of no-self-avoiding polymer conformations with the appropriate Boltzmann weight. This is easily achieved because we already know the total number of random walks of length $p, p-1, p-2, \cdots, 1$ starting from any lattice site i, $(Z_p(i), Z_{p-1}(i), \cdots, Z_1(i)$. We now use this information as follows. First we select the starting position i of a non-self-avoiding random walk of length p with a probability proportional to $Z_p(i)$. Next we consider all neigbours j of site i. The total number of NSAW's of lenght $p-1$ starting from site j is equal to $Z_{p-1}(j)$. We now continue our random walk from i to j with a probability proportional to $Z_{p-1}(j)$. Next we consider all NSAW's of length $p-2$ starting from any of the sites neighboring j, and so on.

It should be stressed that we can enumerate the total number of NSAW's exactly (and cheaply), but that sampling of SAW's is a 'biased' Monte Carlo process that can be more time-consuming.

2.3 Grand-canonical simulations

An interesting application of the moment-propagation scheme is that it allows us to perform simulations of a polymer system *at constant osmotic pressure*, Π. This corresponds to a very common experimental condition. To see how this can be achieved, let us consider the partition function of n ideal polymers of length p in an external field. Clearly,

$$Z_p^{(n)}(\mathbf{r}^M) = [Z_p^{(1)}(\mathbf{r}^M)]^n/n! \,, \tag{9}$$

where $Z^{(1)}$ ($z^{(n)}$) denotes the partition function for one (n) polymer(s). Now let us transform to an ensemble where the polymer chemical potential, μ, is held constant. To be specific, let us again consider the polymer colloid mixture. For convenience, we assume that the total number of colloid particles is held constant. The potential energy function that describes the direct interaction between the colloid particles is denoted by $U(\mathbf{r}^M)$. The partition function for this constant M, V, T, Π-ensemble is:

$$\Xi = \int d\mathbf{r}^M \exp(-U(\mathbf{r}^M)/k_BT) \sum_{n=0}^{\infty} \exp(n\mu/k_BT)[Z_p^{(1)}(\mathbf{r}^M)]^n/n! \,, \tag{10}$$

where we have omitted the constant prefactor. Clearly, the summation in equation 10 yields an exponential,

$$\exp(zZ_p^{(1)}(\mathbf{r}^M)) \,,$$

where we have used the shorthand notation $\exp(\mu/k_BT) \equiv z$. Equation 10 is the starting point for Monte Carlo simulations at constant Π [20]. In such simulations, the relative probability of different colloid configurations is proportional to:

$$P(\mathbf{r}^M) = \exp[-U(\mathbf{r}^M)/k_BT + zZ_p^{(1)}(\mathbf{r}^M)] \,. \tag{11}$$

Note that $Z_p^{(1)}$, which describes the polymer-induced interaction between colloidal particles, can be calculated exactly for every colloid configuration. Of course, such constant-Π simulations are not limited to polymer colloid mixtures. Another interesting application is the simulation of polymer-induced forces between two surfaces. It is worth pointing out that in constant-Π simulations we have no direct knowledge of the actual polymer positions and conformations (although some averages can be computed easily). This demonstrates clearly that great gains in computational efficiency can be made by throwing away 'irrelevant' information.

3 Results

In this section I show two examples that illustrate the power of the moment-propagation method. The first example concerns the observation of self-consistent mode-coupling effects in hydrodynamic long-time tails. The second example shows results of a grand-canonical polymer+colloid simulation. In both cases, the examples are not meant to be anything more than an illustration. For a detailed discussion, the reader is referred to the relevant publications [21, 20].

3.1 Long-time tails

Lattice-gas cellular automata were first introduced as a model for hydrodynamic fluids [22] as a computationally 'cheap' model for a simple fluid. The number of collisions per CPU-second in a lattice gas is at least 4 order-of-magnitude larger than in a hard-sphere fluid. It was therefore hoped that the direct simulation of the tagged-particle motion in a LGCA would yield very accurate data on the algebraic decay of the VACF. In fact, very extensive simulations on two-dimensional LGCA's were performed by several groups [23, 24] with precisely this aim. But the results were disappointing. In fact, the statistics on the long-time tails were poor (typically, an accuracy of 0.3% was the best that could be achieved) and the tails were almost completely swamped in noise. We have applied the moment-propagation method to a model for a lattice-gas on a triangular lattice (to be precise, the 'FHP-III'-model described in ref [13]). Use of the moment-propagation method results in a very drastic reduction of the statistical noise by a factor that can be as much as 10^5 for long times (see figure 2). This noise reduction would correspond to a gain of 10^{10} in computing speed. Admittedly, we have not compared the two techniques on the same computer, but this is not expected to change the picture by more than a factor 10^3 (and probably less). Note that figure 2 shows a very clear non-exponential decay at long times. We can now compare the results of such a simulation with the predictions of lowest order mode-coupling theory (see ref. [25]). Such a comparison is shown in figure 3. The figure clearly shows that the observed tail falls below the predictions of lowest-order mode coupling theory at all densities studied. It should be noted however that the truly asymptotic regime described by eqn. 2 only dominates at times of the order of 10^{20} time-steps. This in inaccessible at present, and

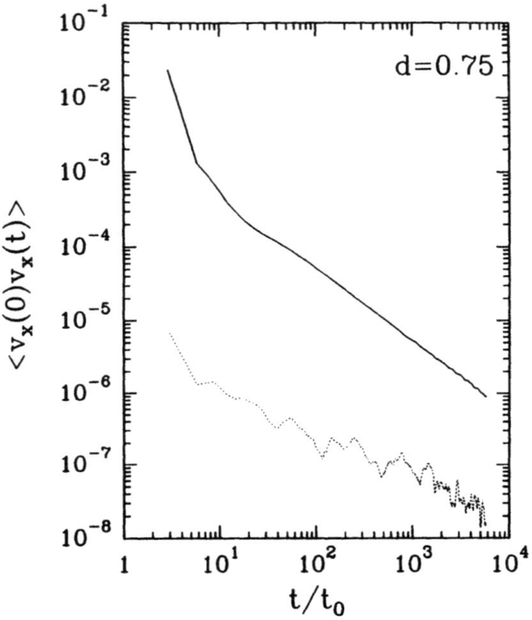

2. Velocity ACF of a tagged particle in a two-dimensional lattice-gas cellular automaton (FHP-III model, see fig. 1). The density d denotes the fractional filling of the lattice. Time is expressed in units of the 'mean-free time' t_0. t_0 is the time that characterizes the initial ('Boltzmann') decay of the VACF. At this density, t_0 equals 0.346 time steps of the lattice gas. The best simulations of the velocity ACF of hard disks extend to some 10^2 mean-free times.

The velocity ACF is indicated by a drawn curve, while the estimated statistical error is shown as a dotted curve. Note that the signal-to-noise ratio hardly deteriorates with increasing time. Note also that the statistical noise at long times is of the order of $2\ 10^{-8}$ which is 5 orders of magnitude better than the best 'conventional' result for the same model system.

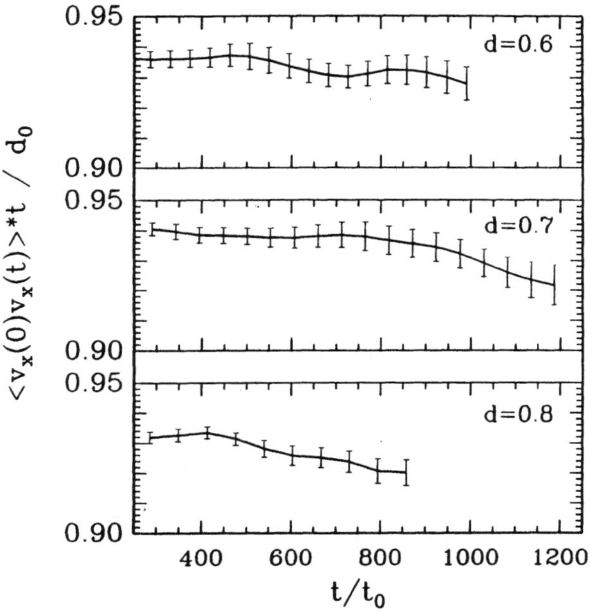

3. This figure shows the long-time tail of a tagged particle in a two-dimensional lattice-gas cellular automaton, divided by the tail predicted by lowest order mode coupling theory. In this theory, the expression for the hydrodynamic tail is of the form d_0/t (see eqn.'1). The explicit expression for d_0 is given in ref. [25]. The density (i.e. the fractional filling of the lattice) is denoted by d. The time is expressed in units t_0 (see fig. 1). For d =0.6,0.7 and 0.8, t_0 equals 0.454, 0.379 and 0.315, respectively.

If the simulation results were adequately described by lowest-order mode-coupling theory, all curves would be constant at a value of 1.0 . However, the figure shows that the velocity ACF is in all cases *lower* than the value predicted by the first-order mode-coupling theory. Moreover, all curves show, on average, a weak, negative slope. Both effects are in agreement with the predictions of self-consistent mode-coupling theory.

almost certainly, also in the future. For more details, the reader is referred to [21].

3.2 Simulations at constant osmotic pressure

The second example is concerned with a well known phenomenon in colloid chemistry, namely that the addition of polymer to a colloidal suspension causes an effective attraction between the colloidal particles. This attraction is purely entropic in origin. It stems from the fact that the total number of allowed polymer conformations is larger when two colloidal particles are close together than when they are far apart. Using the techniques described in section 2.3, it proved possible to observe this effect directly in a computer simulation. Figure 4 shows the radial distribution function, $g(r)$, of a system of three-dimensional hard spheres ('colloidal particles') in contact with a polymer reservoir, such that the polymer 'osmotic pressure' is held constant. For the sake of comparison, the same figure also shows $g(r)$ for the same system in the absence of polymers. Clearly, the addition of polymer to the hard-sphere suspension has resulted in a strong tendency of the spheres to cluster. A more quantitative discussion of this phenomenon can be found in ref. [20]

4 Conclusion

In this paper I have shown how, sometimes, a change of algorithm can make a difference. I have indicated how a numerical technique that was originally devised to study time-correlations in 'sub-molecular' fluids, has been extended to study the structure and phase-behavior of 'supra-molecular' fluids. This opens the way to the numerical study of a class of problems that did, thus far, defy simulation.

Acknowledgements

This work was carried out in collaboration with Martin van der Hoef, Evert Jan Meijer and Germonda Mooij. In addition, I grate-

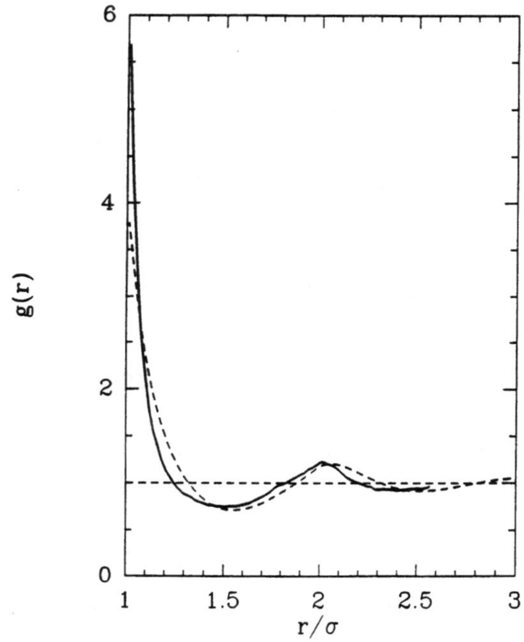

4. Radial distribution function of a system of hard spheres ('colloidal particles') at a reduced density of $\rho^* = 0.56$. The dashed curve shows the result for the pure hard-sphere system. The drawn curve shows the effect on $g(r)$ of addition of polymer.

The present results were obtained for a system of 108 hard spheres in osmotic equilibrium with a bath of chain-molecules with 10 segments. The length of the fully stretched chain-molecule is 5% less than the diameter of the spheres. Hence, these 'polymers' can only cause short-ranged attractions. The osmotic pressure has been chosen such that the average number of polymer molecules in the system is 1.4 10^3.

fully acknowledge the many contributions and suggestions of Mat-thieu Ernst, Henk Lekkerkerker and Ilja Siepmann. The work of the FOM Institute is part of the scientific program of FOM and is supported by the 'Nederlandse Organisatie voor Wetenschappelijk Onderzoek' (NWO)

References

[1] Many early papers on computer simulation can be found in the following reprint collection: G. Ciccotti, D. Frenkel and I. R. McDonald, **Simulation of Liquids and Solids**, North-Holland, Amsterdam, 1987.

[2] B. J. Alder and T. E. Wainwright, Phys. Rev. **A1**:18(1970).

[3] J.R. Dorfman and E.G.D. Cohen, Phys. Rev. Lett. **25**:1257(1970), Phys. Rev. **A6**:776(1972),**A12**:292(1975).

[4] M.H. Ernst, E.H. Hauge and J.M.J. van Leeuwen, Phys. Rev. **A4**:2055(1971).

[5] Y. Pomeau and P. Résibois, Physics Reports **19**:63(1975).

[6] D. Levesque and W.T. Ashurst, Phys. Rev. Lett. **33**:277(1974).

[7] J.J. Erpenbeck and W.W. Wood, Phys. Rev. **A26**:1648(1982).

[8] J.J. Erpenbeck and W.W. Wood, Phys. Rev. **A32**:412(1985).

[9] B.J. Alder and T.E. Wainwright and D. Gass, Phys. Rev. **A4**, 233(1971).

[10] K. Kremer and K. Binder, Computer Physics Reports, **7**:259(1988).

[11] K. Kremer, G. S. Grest and I. Carmesin, Phys. Rev. Lett. **61**:566 (1988)

[12] B. Widom, J. Chem. Phys. **39**:2808 (1963).

[13] D. d'Humières and P. Lallemand, Complex Systems **1**, 599(1987).

[14] U. Frisch, D. d'Humières, B. Hasslacher, P. Lallemand, Y. Pomeau and J.-P. Rivet, Complex Systems **1**, 649(1987).

[15] D. Frenkel in:**Cellular Automata and Modeling of Complex Physical Systems**, P. Manneville, N. Boccara,

G. Y. Vichniac and R. Bidaux (editors), Springer, Berlin, 1989, p.144.

[16] D.Frenkel and M.H.Ernst, Phys. Rev. Lett. **63**:2165(1989).

[17] M.A. van der Hoef and D.Frenkel, Phys. Rev. **A 41** :4277 (1990).

[18] J.I.Siepmann, Mol.Phys. **70**:1145(1990). J.I.Siepmann and D.Frenkel, Mol. Phys. (in press)

[19] G. C. A. M. Mooij and D. Frenkel,Mol. Phys. **74**:41(1991)

[20] E. J. Meijer and D. Frenkel, Phys. Rev. Lett. **67**:1110(1991).

[21] M.A. van der Hoef and D. Frenkel, Phys. Rev. Lett. **66**:1591(1991).

[22] U. Frisch, B. Hasslacher and Y. Pomeau, Phys. Rev. Lett. **56**:1505(1986).

[23] J.-P. Boon and A. Noullez in: Proceedings of workshop on **Discrete kinetic theory, lattice gas dynamics and foundations of hydrodynamics**, World Scientific, Singapore, 1989.

[24] P.M.Binder and D.d'Humières, Los Alamos preprint LA-UR-1341(1988) and P.M. Binder in **Cellular Automata and the modeling of complex physical systems**, P.Manneville, N.Boccara, G.Y.Vichniac and R.Bidaux, editors. Springer, Berlin, 1989, p.155.

[25] M.H.Ernst in: Proceedings of the Les Houches Summerschool on **Liquids, Freezing and the Glass Transition**, J.P.Hansen and D.Levesque, editors, North-Holland, Amsterdam, 1990.

Dynamics of Liquid Water: Fluctuations and Collective Motions

I. Ohmine[1] *and H. Tanaka*[2]

[1]Institute for Molecular Science, Myodaiji, Okazaki, Aichi 444, Japan
[2]Department of Molecular Engineering, Kyoto University, Sakyoku, Kyoto 606, Japan

Abstract. Intermolecular binding (hydrogen bonds) structure reorganization in liquid water involves collective displacements of water molecules, exhibiting large energy fluctuations. The time scale, spatial scale and energy scale of these collective motions are analyzed by using the molecular dynamics, quenching and mode excitation methods.

1. Introduction

Liquid water is a frustrated system and has many fundamental binding structures due to the strong directionality and large cohesion energies of hydrogen bonds. Upon melting from ice to liquid, water absorbs a large latent heat and the hydrogen bonds between water molecules are partially broken and frustrated. The system then starts to undergo transitions from one fundamental structure to another, inducing collective motions and large energy fluctuations. Liquid water thus shows interesting physical and chemical properties, and, as the most commonly used solvent, yields significant solvent effects on chemical reaction dynamics.

We have analyzed the dynamical behavior of liquid water in terms of the potential energy surface involved and have examined longtime scale fluctuations of liquid water. Potential energy surfaces of liquids consist of many energy wells. Each well is represented by its energy minimum, called an inherent structure (a fundamental binding structure) [1]. We use the inherent structures successively visited by the system to determine molecular displacements and activation energy barriers in the water structural change. We first analyze the potential energy surface along a trajectory [2]. Degrees of the molecular displacement involved in inherent structure transitions are determined and grouping of the successive inherent structures is made. The transition between the inherent structure groups involves large collective motions [3]. Reaction coordinates connecting the inherent structures and their energy barrier height are determined [4]. We then studied the multidimensional nature of the potential energy surface by performing mode excitation dynamics and analyzing the mechanism of the energy relaxation process through mode mixing [5].

It is, of course, not sufficient to investigate the water dynamics only by means of potential energy surface analysis. The dynamics proceeds significantly above the potential energy barriers and so the question is how much it picks up the potential energy information. The liquid dynamics, involving multiminimum potential energy surfaces with frustrated interactions, has a strong relation with general dynamical problems such as spin-glass dynamics [6], 1/f fluctuations [7], chaos [8] and many others. A discussion of the water dynamical behavior related these dynamics will be given. The studies particularly related to the present one are by Berry and his coworkers investigating potential energy and dynamics of small clusters near the 'phase transition' point [9-10].

2 Energy fluctuations

A trajectory calculation is performed, and then successive water structures in certain intervals in the trajectory are quenched to their local minima (inherent structures) [1]. In this way we obtained successive inherent structures, representing the fundamental structures, along the trajectory [3]. An inherent structure is a geometry corresponding to a local minimum of the total system potential energy. Individual intermolecular interactions are not necessarily optimized and can even be repulsive in an inherent structure; a few individual water molecule potential energies are very unstable. The energy difference between the most stable and unstable water molecules can reach up to 20 kcal/mol [2]. The stability and instability of individual molecules alter with time, and a very large potential energy fluctuation, 15-20 kcal/mol, occurs in 10 ps for each molecule. See-saw type energy exchanges among neighbor molecules are observed [3].

There are various configurations of the inherent structures and their energies (the system total energies of these structures) also fluctuate, as seen in Fig.1. This energy fluctuation consists of very fast (10-15 fs), medium and slow (seems to be a few ps) components [4].

It is noted that this energy fluctuation is very similar to that found in the sand pile model of the so called self-organizing criticality problem [7]. We have recently found that the

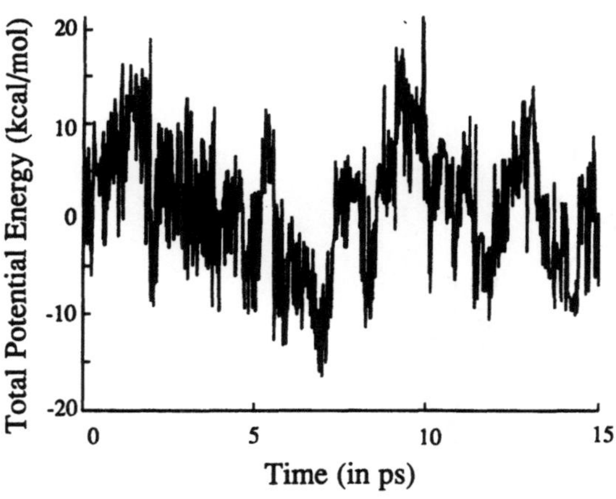

Figure 1. The total potential energies of the system with 64 water molecules in the inherent structures. The 15 ps trajectory with T=297K is quenched in each 10 fs interval. Energy is kcal/mol and relative to an arbitrarily chosen energy as zero. Time in ps. See also Ref.[4].

total potential energy fluctuation of liquid water yields a near $1/f^{\alpha}$ (α=0.7-0.8) power spectrum for ω=1-1000cm^{-1}, indicating that various relaxation processes, especially long time scale reorganization of water binding structure, are involved in the water dynamics. This spectrum is the reflection of the fundamental structural energy change shown in Fig.1. This water spectrum is in strong contrast to the white noise spectrum of liquid argon. A more detailed discussion on the $1/f^{\alpha}$ noise will be given elsewhere [11]. This fact also indicates that very long time (a few tens of ps) MD steps or very many MC steps are needed to simulate the water dynamics, especially to deal with properties related to its binding structure reorganization.

3. Collective motions

In the analysis of inherent structures, it became clear that the water dynamics mostly makes successive small (not so distinctive) structural changes, which occasionally yields large distinctive structural reorganization [4]. The small changes are always possible without inducing much constraint. But when the constraint reaches a threshold, the system must induce large reorganization due to the strong directionality of the hydrogen bonds. The system with 64 water molecules was found to show a clear group structure (The system with 216 water molecules yields a very similar but a slightly less clear group structure, since it involves more flexible translational motions.) Some sequential inherent structures, though they are distinct in the strict sense, yield almost the same geometry and so form a group [4]. The transitions between these groups of inherent structures are very distinct. This grouping of inherent structures is schematically represented in Fig.2.

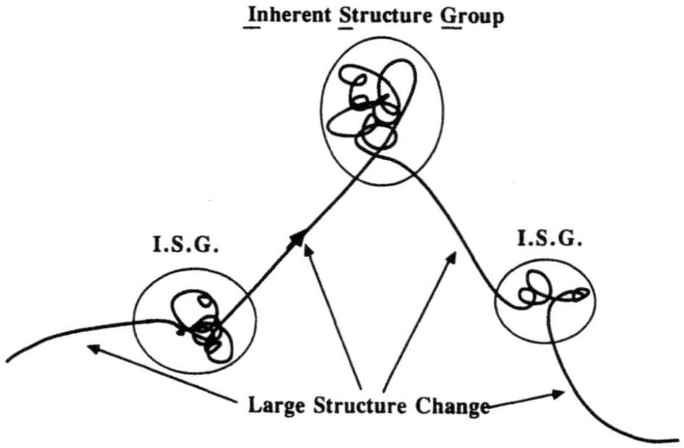

Figure 2. Schematic representation of the inherent structure transitions successively occuring in liquid water dynamics. Many successive inherent structures are very similar and thus form a group. Large displacements take place between the groups of the inherent structures.

4. Reaction coordinates and activation energies

To understand the mechanism of transitions between inherent structures, it is important to know transition state (TS) geometries and energies of reaction coordinates connecting inherent structures. It is, however, a very difficult task to determine the TSs for systems with many degrees of freedom. Recently, Wales et al. [12] have modified a 'walk uphill' method, using the second derivative matrix (Hessian matrix) of the potential energy surface, proposed by Cerjan and Miller [13] and applied to argon cluster systems to investigate their dynamics and 'phase transition'. We have also applied a TS search method, using the Hessian matrix but including the higher order correction to the Cerjan-Miller method (correction optimizing the gradient for the coordinates orthogonal to the 'reaction coordinate' being zero) to small cluster systems and have found many new TSs. This method, called the gradient extremal method, was developed by Ruedenberg et al. [14] and applied for several molecular systems by Shida et al.[15] and others.

For water, however, we have only used a very approximate method developed by Elber and Karplus [16]. This method is convenient (efficient) to briefly estimate the energy barrier height and molecular movements along the reaction coordinates for systems with very many degrees of freedom. The method is to minimize the average potential energy along a reaction path, and thus it gives only an approximate result. It was found [4] that molecular motions are strongly correlated and so barrier energies for reaction coordinates are about 1/3 of those for least motion paths (Fig.3). The distribution of the transition state energies for 54 arbitrarily chosen inherent structure pairs is listed in Table 1. The activation mechanism with the average activation energy, about 3 kcal/mol, predicts that the inherent structure transitions occur once about every 10 fs for the 64 molecular system. This value is close to the transition frequency observed in the trajectory calculation, 600 transitions in 8 ps. The energy barriers for distinct transitions between successive groups of inherent structures are found to be also small, a few kcal/mol. This suggests that occurrence of the distinct transitions is restricted not by activation but rather by narrow bottlenecks in phase space.

Table 1. Transition state energy distribution for the system with 64 water molecules, and 54 arbitrarly chosen pairs of successive inherent structures. See also Ref.[4].

Barrier Energy	Number of Transition States
0-2 kcal/mol	15
2-4	18
4-6	16
6-8	4
8-10	1

There are, however, some barriers reaching up to 20 kcal/mol for certain inherent structure pairs (Fig.4) [4]. This occurs when the system goes through ridges of deep minima. The trajectory must proceed without being trapped in such deep minima, i.e. it is ballistic. Then the question arises as to how fast the vibrational mode relaxation is in liquid water. The vibra-

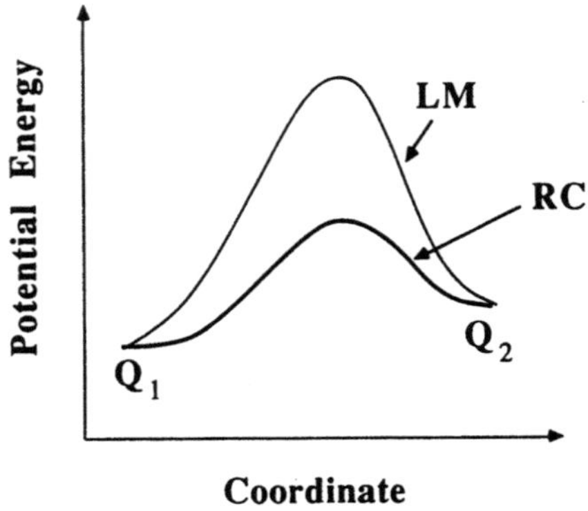

Figure 3. Total potential energies along the reaction coordinate (RC) and along the least motion path (LM) connecting two inherent structures Q_1 and Q_2. LM is the shortest path (in Euler angles for rotations of the molecules and in Cartesian coordinate for the center of mass motions) connecting these inherent structures. See also Ref.[4].

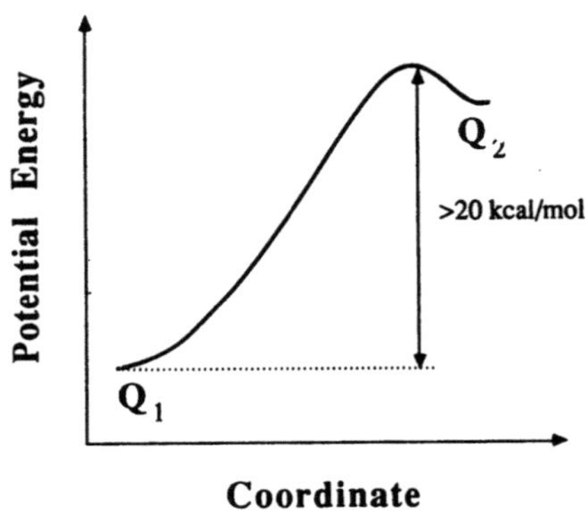

Figure 4. The large potential energy barriers sometimes observed in water dynamics.

tional relaxation is slow for some modes. But how about for the other modes? This point is explored in terms of the normal mode excitation dynamics (see below). We expect that a structured liquid such as liquid water must behave as being in between ballistic and activation dynamics in the molecular dynamical time scale (from subps to ns).

5. Mode excitation: the multidimensional nature of potential energy surface

The normal modes for liquid water (at the inherent structures) have been analyzed by Pohorille et al. [17] and by us [3,4]. The normal mode frequency distribution is schematically represented in Fig.5. High frequency components (450-1000cm^{-1}) are

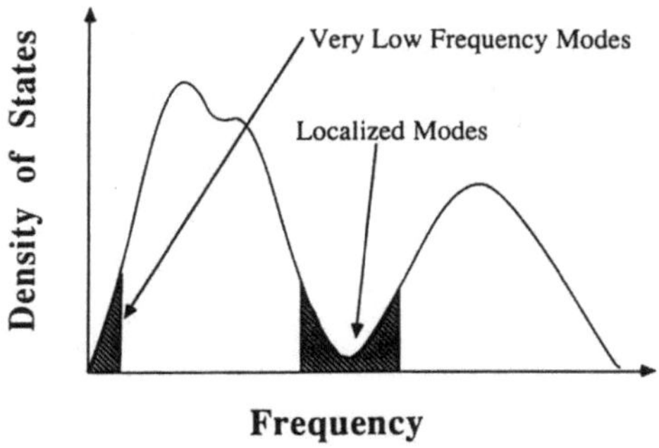

Figure 5. Normal mode distribution. Schematic representation.

the librational modes and low frequency components (10-280cm^{-1}) are the translational modes. In between these components (280-450cm^{-1}) there are very localized modes, consisting mainly of translations mixed with some librations [3,4,17].

To explore the multidimensional nature of the potential energy surface (not only along the trajectory), we performed trajectory calculations initially exciting certain normal modes, and then analyzed the mode energy relaxation process [5]. We have calculated many normal mode excitation dynamics: different trajectories for different normal mode excitations, for different inherent structures, and for different excitation energies. It was found that most modes individually couple only with one or two other modes, and thus exhibit near recurrence of mode energies in a few tens of picoseconds (Fig.6). Only localized modes (Fig.6) and the very lowest frequency modes yield fast energy decay. These modes couple with double frequency modes through the 1:2 Fermi resonance, as well as with nearby frequency modes. When the mode excitation energies for

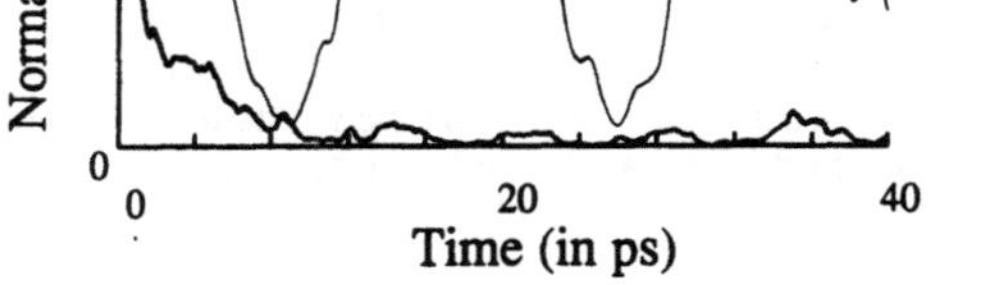

Figure 6. The mode energy decay for a high frequency mode excitation (10th mode) and for an intermediate frequency localized mode excitation (188th mode). The system is initially excited to the direction of a certain mode (i=10 or 188) and the energy of that mode is monitored and plotted. The mode energy decays since the energy flows into other modes through mode mixing. Energies are normalized to the initial excitation energy of 1. Time in ps. See also Ref[5].

Figure 7. Schematic representation of the bifurcation of trajectories leading to different inherent structure transitions on the (two dimensional) potential energy surface of the total system. Different excitation energies of the same normal mode result in the different inherent structure transitions.

these modes exceed certain threshold values, inherent structure transitions are induced. The threshold energies for the lowest frequency modes are very low, and can be as low as 0.05 kcal/mol. The inherent structure transitions induced by such small excitation energies are usually very localized; some single molecules can dislocate in the water binding structure without a significant energy barrier, by changing their binding partners. As the excitation energies increase, the excitation of a low frequency mode results in a different inherent structure transition, involving collective displacements of many molecules. This bifurcation of the reaction coordinates (Fig.7) to different inherent structures is the origin of the rich variety of water dynamics.

Most modes, as we have found, yield very slow relaxation and if only these 'inactive' modes are involved in the dynamics, the dynamical process is said to be ballistic. If the 'active' (i.e. fast decaying) localized and low frequency modes are involved, the system may be trapped in a local potential well and the process becomes an activation type [4].

6. 'Dynamics': problems to be solved

So far, we have tried to understand the dynamical properties of liquid water mainly through analyses of the potential energy surface involved. This approach has inevitable limitations, since liquid systems have usually much higher energy than most of their energy barriers. So, the systems do not see the details of potential energy surfaces. If it does not see even the global features of the potential energy surface, the system behaves as simple diffusion. Liquid water, however, is highly structured due to the strong hydrogen bonds, and so its dynamics must be strongly correlated to general features of the potential energy surface. Besides, even for simple liquids (clusters), the influence of the potential energy surface on the dynamics should become prominent when the temperature approaches melting point (at phase transition). Berry et al. [10] have recently calculated the Kolmogorov (K) entropy for a small cluster system, Ar_3, examined how the trajectory covers the phase space and analyzed its temperature dependence. They found that K entropy yields singular behavior when the transitions over energy barriers start taking place. The system becomes chaotic long before these inherent structure transitions occur. This type of analysis of phase space dynamics will be also very useful to understand liquid water dynamics; we need to go beyond the K entropy and find a quantity more directly describing the trajectory development on the multi-minimum potential energy surface. Aizawa and coworkers show that the power spectrum of dynamical fluctuations is of $1/f^{\alpha}$ type if the dynamics proceeds awhile on a torus in phase space and then makes a quick jump to proceed on another torus and so on [18]; this situation might be very similar to the dynamics of structure liquids.

The fast and large fluctuation in liquid water occurs as a mixture of collective behavior near the phase transition at the lower temperature and independent behavior at the higher temperature. It is thus interesting to see how the time-, energy- and spatial-scales of the fluctuation alter with temperature. Our preliminary study [11] shows that the power spectrum of the potential energy fluctuation correlation yields $1/f^{\alpha}$ type fluctuation, suggesting the existence of long time correlation in fluctuations. The half decay time of this correlation function significantly increases and the spectrum significantly changes as the temperature goes down. It will be very important to analyze how the correlation time diverges as the temperature approaches a phase transition point [19]. It is very useful to make use of the characterization of the hydrogen bond network and its restructuring pathway in order to classify the low temperature dynamics near the spinodal point. The studies by Kataoka et al. [20], Stanley et al. [21] and Sasai [22] have explored such a direction. These type of studies are to be combined with the phase space theory to obtain a penetrating view of the dynamics of structure liquids, such as water.

Acknowledgements

We thank Dr. M.Sasai and Dr. N.Shida, and Prof. P.G.Wolynes and Prof. R.Ramaswamy for cooperative work and stimulating discussions. The work is supported partially by a Grant-in-Aid for Scientific Research in the Priority Area of 'Molecular Approaches to Non-equilibrium Processes in Solutions' from the Ministry of Education, Science and Culture, Japan.

References

1) H.F.Stillinger and T.A.Weber, Phys.Rev.25,978(1982); 28,2408(1983); J.Phys.Chem.87,2833(1983).
2) H.Tanaka and I.Ohmine, J.Chem.Phys.87,6128(1987).

3) I.Ohmine, H.Tanaka and P.G.Wolynes, J.Chem.Phys.89,5852(1988).
4) H.Tanaka and I.Ohmine, J.Chem.Phys.91,6318(1989).
5) I.Ohmine and H.Tanaka, J.Chem.Phys.93,xxxx(1990) in press.
6) D.Chowdhury, Spin Glass and Other Frustrated Systems (World Science,Singapore,1986).
7) P.Bak, C.Tang and K.Wiesenfeld, Phys.Rev.Letters,59,381 (1987);P.Dutta and P.M.Horn, Rev.Mod.Phys.53,497(1981); M.B.Weissman, Rev.Mod.Phys.60,537(1989).
8) G.M.Zaslavshy, Chaos in Dynamic Systems(Harwood Academic Publishers,Chur,Switzerland,1984).
9) F.Armar and R.S.Berry, J.Chem.Phys.85,5943(1986).
10) T.L.Beck, D.M.Leitner and R.S.Berry, J.Chem.Phys.89,1681(1988).
11) M.Sasai, I.Ohmine and R.Ramaswamy (in preparation).
12) D.J.Wales and R.S.Berry, J.Chem.Phys.92,4283(1990); H.L.Davis, D.J.Wales and R.S.Berry,ibid.,92,4308(1989); D.J.Wales, ibid.91,7002(1989); Chem.Phys.Letters, 166,419(1990).
13) C.J.Cerjan and W.H.Miller, J.Chem.Phys.65,2800(1981).
14) D.K.Hoffman, R.S.Nord and K.Ruedenberg, Theor.Chim.Acta, 69,265(1986); P.Valtazanos and K.Ruedenberg,ibid. 69,281(1986).
15) N.Shida, J.E.Almlof and P.F.Barbara, ibid.76,7(1989).
16) R.Elber and M.Karplus, Chem.Phys.Letters,139,375(1987).
17) A.Pohorille, L.R.Pratt, R.A.LaViolette, M.A.Wilson and R.D.MacElory, J.Chem.Phys.87,6070(1987).
18) Y.Aizawa, Y.Kikuchi, T.Harayama, K.Yamamoto, M.Ota and T.Tanaka, Proc.Theor.Phys.Suppl,985,36(1989).
19) C.A.Angel, Water, a Comprehensive Treaties, edited by F.Franks (Plenum, New York,1982); R.J.Speedy and C.A.Angel, J,Chem.Phys.65,851(1976).
20) Y.Kataoka, H.Hamada, S.Nose and T.Yamamoto, J.Chem.Phys. 77,5699(1982).
21) H.E.Stanley and T,Teixeira, J.Chem.Phys,73,3404,(1980); R.L.Blumberg, H.E.Stanley, G.A.Geiger and P.Mausbach,ibid. 80,5230(1984); F.Sciortino, P.H.Poole, H.E.Stanley and S.Havlin, Phys.Rev.Letters,64,1686(1990).
22) M.Sasai, J.Chem.Phys.93,7326,(1990).

Ab Initio Hydrodynamics via Atomistic Simulation

D.C. Rapaport

Physics Department, Bar-Ilan University, Ramat-Gan, Israel

Abstract This lecture addresses the modeling of hydrodynamic phenomena in a manner that is free from assumptions concerning the properties of bulk fluids, namely by allowing the flow to emerge as the natural response of a microscopic system of interacting particles to an appropriate driving force. Complex flow phenomena which this approach has succeeded in reproducing include thermal convection and flow past an obstacle. Large numbers of particles are generally required for such simulations, and specialized techniques have been developed to allow the computations to utilize advanced vector and distributed processor systems. The computational methods and recent results are described.

1. Background

The underlying equations of continuum hydrodynamics are based on the laws of conservation of matter, momentum and energy, supplemented by a series of empirical constitutive relations that embody the detailed transport properties of the fluid [1]. While there can be little doubt that these equations are extremely accurate and versatile in dealing with a wide variety of flow problems, at no stage does this formalism acknowledge the atomistic nature of matter. This can lead to problems when the nature of the transport process is sufficiently complicated that it can no longer be described in terms of simple constitutive relations; examples of fluids where deviations from the simple description are substantial include low-density gases in which ballistic motion is important, and polymer fluids where chain entanglement can generate viscoelasticity.

Molecular dynamics addresses the general problem of classical many-body systems at the atomistic level. Given the constituent particles and their interactions the equations of motion are integrated in complete detail [2]. The hydrodynamic fields that correspond to macroscopic observables are evaluated by averaging both over the particles contained in small spatial regions and over short time intervals. This coarse-graining represents an enormous reduction in the amount of information characterizing the behavior; while all details at the level of individual particles are eventually discarded, the resulting fields are only available by virtue of the fact that particle trajectories have been computed in full detail – satisfactory shortcuts for reducing the complexity of this intermediate step (along the lines of cellular automata [3], but based on a microscopic picture closer to reality) have yet to be devised.

2. Methods

2.1 Basic molecular dynamics

Molecular dynamics simulation amounts to solving the classical many-body problem in any of a wide variety of physical contexts. Some simulations require only modest amounts of computation, while others can easily consume literally hundreds of hours on the fastest computers. For most applications [4] it has proved perfectly adequate to consider systems containing between several hundred and several thousand atoms (or molecules). There is, however, no shortage of problems for which much larger systems are required; one example – the subject of this lecture – is the modeling of fluid flow at the atomistic level. In studies of this kind there are certain natural length scales inherent in the flow patterns which emerge spontaneously during the simulation; very large numbers of atoms (as well as runs of long duration) are necessary in order to provide the space (and time) required for this pattern development to proceed unhindered. While there are a number of techniques that have been developed to enhance the efficiency of the computations, the increasing reliance on vector and parallel processing has led to algorithms that are tailored to individual processor architectures.

At this early stage in the application of molecular dynamics methods to the study of fluid flow, the focus is exclusively on problems in which the interatomic forces are of very short range, typically little larger than the diameter of the atoms themselves. This proves adequate for studying the most prominent features of several classes of flow phenomena. Such a limited interaction range results in only very few interaction partners per atom even in a dense fluid, and solving the dynamics of this kind of system will obviously require substantially less effort than the corresponding problem with interactions of longer range. However, there is a need for an effective bookkeeping scheme that maintains a record of which atoms lie within mutual interaction range; in the absence of such a scheme, the fact that the spatial configuration of the fluid is continually changing would result in a computation whose complexity grows as the square of the number of atoms [2].

The most obvious bookkeeping scheme is based on subdividing the system into a set of relatively small cells. If the edges of the cells exceed the interaction range then interactions can only involve pairs of atoms in either the same or immediately adjacent cells. Since the determination of cell membership for each atom requires a fixed computational effort, and the interaction computations address a neighborhood whose size is independent of the total number of atoms, the computational effort is proportional to the number of atoms in the system. Linked lists provide an economical method for representing cell occupancy data.

The use of neighbor lists leads to a further reduction in computational effort, with the aforementioned cells being used to ensure that neighbor list construction proceeds efficiently. If a list of all the atoms present in a spherical (circular in two dimensions) neighborhood of a given atom is prepared, then if the neighborhood size is marginally larger than the interaction range it is apparent that the list will include all interaction partners of that atom for several subsequent time steps. The gain in performance over the use of cells depends on the ratio of the volume (or area) of the neighborhood to the combined volume (area) of all the relevant cells. The atomic displacements can be monitored to determine when the neighbor lists should be rebuilt; this must occur no later than the earliest instant at which an atom not originally in the neighborhood could come within interaction range. There is a tradeoff between storage and computation speed involved in this approach, and the storage requirements can easily double if neighbor lists are used; given that high performance computer memory is still relatively costly this additional storage is not always available.

While the above methods have proved highly effective on scalar processors [2], the question is how to adapt them for use in vector and distributed processing environments.

2.2 Vector processing

Algorithms that vectorize effectively are characterized by the particular manner in which data is accessed and the the kind of processing involved. The reason that certain modes of operation are preferred is that the processor handles both memory access and arithmetic in a pipelined mode, with a resulting throughput substantially higher than what would be possible if each operation were to be carried out separately without overlap [5]. Pipelining is only possible if the same operation is performed repeatedly on a set of data items arranged in a systematic way, and deviations from the patterns dictated by the hardware result in seriously degraded performance.

The layer method of reorganizing cell data leads to a vectorizable computation which retains the benefits of the cell approach [6]. Instead of representing cell occupancy using linked lists, the identities of the cell occupants are placed in a set of arrays: each such array, or layer, contains space for one atom per cell, and the total number of arrays must not be less than the maximum possible cell occupancy. The entire interaction computation then consists of pairing layers using all allowed offsets between adjacent cells, with a vectorizable innermost loop that processes pairs of atoms specified in the layers. Conditional operations are incorporated in the vectorized loops to ensure that the calculations are actually carried out only if both layer positions specify valid atoms.

Two vectorized forms of the layered interaction computation have been constructed to match the specific hardware features of different processor families. One version represents the cells requiring attention in each layer by means of bit vectors, the other employs sets of indices to represent layer contents. Both approaches lead to fully vectorized computations, but the latter is inefficient from the storage point of view; this objection can be overcome by partitioning the system spatially and treating each of the subsystems separately, in a manner reminiscent of the approach (described below) used for distributed processing. The layer method can also be applied to neighbor lists.

2.3 Distributed processing

There are two ways to implement a molecular dynamics calculation in a multiprocessing environment based on distributed (as opposed to shared) memory and message passing. The first assigns the atoms permanently to particular processors, with the consequence that the communication load is proportional to the total number of atoms. The second is based on spatial subdivision, with subregions of the simulation volume assigned to different processors and atoms allowed to migrate between processors as necessary; for short-ranged interactions the communication load depends primarily on the number of atoms close to subregion boundaries [6]. The latter scheme, with its greatly reduced dependence on communication, is clearly more efficient in this case.

The computations for the separate subregions are carried out concurrently. Message passing between processors responsible for adjacent subregions is required at each time step, both to permit the computation of interactions between atoms close to, but on opposite sites of mutual subregion boundaries, as well as to allow migration between subregions. Since (for sufficiently large systems) only a small fraction of the atoms are close to subregion boundaries, the total communication time is small compared to the computation time.

The most recent generation of superscalar processors have established themselves as cost-effective competitors of the vector processor; they have the additional advantage of not making the same stringent demands on the structure of the computations. An enormous potential exists for using arrays of such devices in order to achieve extremely high performance at tolerable cost. While the simplest multiprocessor design involves distributed memory, with data sharing being achieved through message passing as described above, the alternative shared memory approach might lead to simpler algorithms, although in this case the hardware (in particular the capability of dealing with memory access conflicts) is more complex. There can be little doubt that irrespective of the architecture of the individual processor a multiplicity of processors will yield substantial performance improvements across a wide range of applications – ideally proportional to the number of processors involved.

2.4 Step potentials

An alternative to the continuous potentials implicit in the preceding discussion is the use of models based on step potentials. The two most familiar cases are the repulsive hard sphere (or disk) and the square-well fluids, with easy extension to more complex situations such as polymer chains. Entirely different techniques for solving the equations of motion are required due to the fact that the potential is not continuous.

The particles of the fluid move in straight lines except when involved in collisions. Collision events are instantaneous, and formulae exist for the velocity changes given the incoming state [2]. Determining if and when a collision between two particles occurs is also a simple exercise in algebra. The simulation does not progress in a series of constant time steps, but as a succession of irregularly spaced events that correspond to collisions. The problem with this approach arises from the need to handle large amounts of data concerning possible future collisions and the consequent effort of determining which will be the next to occur [7]. Since it is impossible to examine all potential collisions in order to choose the earliest, it is necessary to maintain a calendar of potential future collisions and make minor alterations after each collision occurs (namely the addition and deletion of potential collisions for the particles involved); the only occasion all possible collisions need to be examined is when the calendar is initially constructed.

The two salient features of an efficient implementation of a hard sphere simulation are the use of a cell subdivision analogous to that introduced earlier, and the representation of the event calendar by means of a data structure organized as a binary tree. The use of cells ensures that the only potential collisions that are examined are those bet veen particles that are already relatively close to one another; in order to ensure that no collisions are omitted it is necessary to introduce additional events corresponding to cell occupancy changes (i.e., particles crossing cell walls). The use of a binary tree for organizing the event calendar ensures that only minimal processing is required for both updating the calendar and locating the next event (the operations required for dealing with the tree depend logarithmically on particle number). The nature of the computations used for step potentials render the approach unsuitable for vectorization, and although a distributed approach is possible in principle, the details have yet to be explored.

3. Flow simulations

3.1 Atomistic hydrodynamics

As pointed out in the introductory section continuum hydrodynamics does not acknowledge the fact that fluids consist of discrete molecules (the theory even predates the universal acceptance of the atomic nature of matter). The initial motivation for studying flow problems by means of molecular dynamics was one of curiosity: can a fluid consisting of a comparatively small number of simple particles exhibit behavior characteristic of macroscopic flows? It should not be forgotten that even the largest simulations involve submicroscpic length scales only, and thus it is not *a priori* obvious that any strong resemblance should emerge. Some flow situations are more readily realized than others, and this is reflected in the choice of case studies. The two-dimensional nature of these systems also contributes to their attractiveness.

3.2 Rayleigh-Bénard convection

The Rayleigh-Bénard problem [8] deals with a horizontal fluid layer between two plates maintained at different temperatures – the lower plate is the hotter one. The opposing effects of buoyancy and gravity can lead to the appearance of convective cells that, under suitable conditions, take the form of linear rolls. In addition to stationary patterns the system can be made to exhibit periodic time-dependence and even chaos. The behavior is governed by two dimensionless quantities – the Rayleigh and Prandtl numbers – with the ratio of the lengths of the sides of the container as well as the thermal properties of the walls also playing a role.

In order to model convection by means of molecular dynamics [9,10] a two-dimensional container containing a hard-disk fluid is studied. The thermal effects of the hot and cold walls are reproduced by requiring each particle that collides with a wall to be reflected with a velocity corresponding to the wall temperature; in addition, the walls can be of slip or non-slip type, a feature also readily included in the boundary collision prescription. Laterally, the system can have periodic boundaries, or hard walls that are again either slip or non-slip. Gravity is incorporated by requiring the particle trajectories between collisions to be parabolic rather than linear; the immediate outcomes of collision events and the prediction of future collisions are not altered by the presence of gravity. One reason for choosing hard disks as the constituents of the fluid, rather than soft particles, is the fact that the temperature gradient would require that the constant time step in the latter case be determined by the hottest part of the fluid, with a consequent waste of computational effort for the colder region; for hard disks the collision events adjust the effective local "time step" automatically. Establishing rigid boundaries also tends to be simpler with hard disks.

A variety of different forms of behavior have been observed, depending on the values of the parameters defining the system. The evolution of the flow is also found to depend in an unpredictable manner on the initial conditions, as is the case experimentally; the differences in evolution can vary from minor details to entirely different flow patterns. The kinds of pattern development that can be seen include the possibility of the fluid remaining in a non-convective state (no persistent patterns, although convective structures may appear as transients while the system achieves its conducting steady state), convergence to a particular roll pattern which itself may be metastable and eventually converted into yet another roll pattern, and, the most recent observed results of apparently periodic roll rearrangement [10]. Examples of these results are shown in Figure 1. The larger systems – containing as many as 10^5 disks – already require several hundred hours of mainframe computer time for a run of barely adequate duration, and while it is clearly desirable to make a more thorough coverage of parameter space than has been accomplished so far, progress with the current generation of computers is still painfully slow.

3.3 Obstructed flow

When a fluid flows past a bluff (non-streamlined) obstacle, even at low velocity where the flow remains laminar, prominent structures appear in the downstream wake [1,8]. Depending on the value of the dimensionless Reynolds number, behavior ranging from stationary eddies to

Figure 1. Periodic convective rolls; the plots show uniformly spaced contours of the stream function.

time-dependent periodic vortex patterns can be observed; turbulence is also possible under the appropriate conditions, but this lies beyond the scope of the present work.

Once again the simulation addresses a two-dimensional problem, but now using soft disks. These are more economical in terms of data storage – since the simulations have involved up to nearly 3×10^5 particles this is an important consideration.

The computations are based on a standard molecular dynamics approach – using the various computational enhancements described earlier where possible – modified to allow both the inclusion of a circular obstacle and a mechanism for driving the flow. The obstacle boundary can be of either slip or non-slip type. Both cases can be implemented by means of a suitably defined interaction term: for a slip boundary this corresponds to a repulsive radial force which diverges close to the obstacle periphery, whereas the effect of a non-slip boundary can be created by a ring of closely-spaced infinitely massive soft disks that are otherwise identical to the fluid particles. Fluid flow is enforced either by means of a uniform gravitational field or with a pressure head created by periodically altering the velocities of the particles near the flow inlet to give them a downstream bias. In addition, the particles near the inlet must also have their velocities adjusted to maintain a constant temperature in the system (which would otherwise increase monotonically owing to viscous heating) as well as to ensure that any memory of the downstream flow patterns which might be reintroduced upstream due to the use of periodic boundary conditions is eliminated.

With the appropriate choice of parameters (obstacle size and flow rate) it is possible to observe the formation of a pair of recirculating eddies at the downstream edge of the obstacle during the initial stages of flow. As the systems considered have become progressively larger it has proved

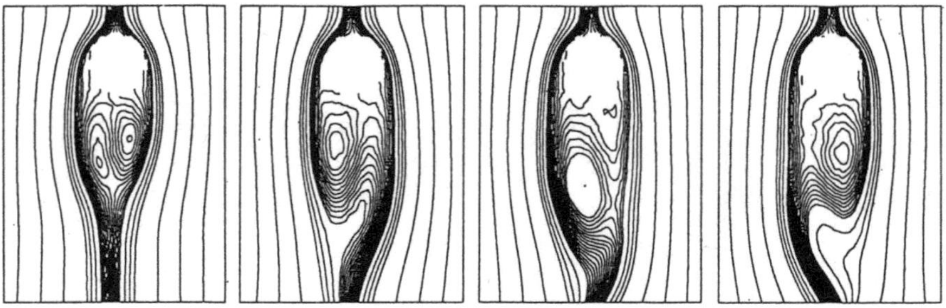

Figure 2. Obstructed flow – eddy development and evolution of an oscillatory wake; the stream function contours are more closely spaced near the center of the flow to show wake structure in greater detail.

possible to differentiate between the case of a confined, almost stationary wake envelope, and an extended oscillating wake that corresponds to the familiar vortex street [11]. The need for extremely large systems stems from the fact that the downstream boundary of the system can interfere with the the natural development of the wake structure. Typical flow patterns are illustrated in Figure 2.

4. Future prospects

The present studies are only early steps along this novel path to flow exploration. While the basic concept of modeling systems at the most elementary level, with the goal of allowing spontaneous cooperativity to emerge naturally, dates back at least as far as Laplace, it is only in recent years that computer power has made such an approach viable. Present computer technology allows just a glimpse of what will be achievable once hardware speeds and storage capacity increase by the expected orders of magnitude based on extrapolation of past growth.

The eventual purpose of these studies is not to compete with the engineering approach based on the solution of the equations of continuum hydrodynamics (the latter having a clear advantage when applicable, and also a beneficiary of future advances in computer hardware), but rather to facilitate an understanding of the connection between the microscopic motions and correlations (both the spatial and temporal kinds) on the one hand and the observed macroscopic behavior of the fluid on the other. There are many instances of having to resort to phenomenology in order to produce engineering solutions, and it is in such circumstances that the molecular dynamics approach, with its ability to ask meaningful questions at the atomistic level, provides possibly the only method of achieving an understanding of the mechanisms involved. One can well envisage the results of such microscale studies leading to an appreciation of why the engineering phenomenology succeeds, as well as to more precise empirical rules in cases where the engineering approach has yet to demonstrate its reliability.

References

[1] G. K. Batchelor, *An Introduction to Fluid Dynamics* (Cambridge University Press, Cambridge, 1967).

[2] M. P. Allen and D. J. Tildesley, *Computer Simulation of Liquids* (Oxford University Press, Oxford, 1987).

[3] S. Wolfram, *Theory and Applications of Cellular Automata* (World Scientific, Singapore, 1986).

[4] *Molecular Dynamics Simulation of Statistical Mechanical Systems*, eds. G. Ciccotti and W. G. Hoover (North-Holland, 1986).

[5] R. W. Hockney and C. R. Jesshope, *Parallel Computers*, 2nd edn., (Adam Hilger, Bristol, 1988).
[6] D. C. Rapaport, Computer Phys. Repts. **9**, 1 (1988), and Computer Phys. Commun. (in press).
[7] D. C. Rapaport, J. Comput. Phys. **34**, 184 (1980).
[8] D. J. Tritton, *Physical Fluid Dynamics* 2nd edn., (Oxford University Press, Oxford, 1988).
[9] M. Mareschal and E. Kestemont, J. Stat. Phys. **48**, 1187 (1987).
[10] D. C. Rapaport, Phys. Rev. Lett. **60**, 2480 (1988), and to be published.
[11] D. C. Rapaport, Phys. Rev. A **36**, 3288 (1987), and to be published.

Part V

Ab Initio Molecular Dynamics Simulations

Ab Initio Molecular Dynamics

M. Parrinello

IBM Research Division, Zurich Research Laboratory,
CH-8803 Rüschlikon, Switzerland

Abstract. A recently developed method allows us to perform molecular dynamics simulations where the interatomic forces are at any instant of time derived from first-principle electronic calculations, in the framework of density functional theory. This scheme has allowed the solution of real material problems which until a few years ago had been considered as inaccessible. Selected applications will be presented, in particular, the high-pressure region of the carbon phase diagram and some properties of I-IV liquid alloy.

1. Introduction

A crucial issue in molecular dynamics (MD) simulation of the behavior of materials is a realistic modelling of interatomic forces. The reliability and predictive power of the calculation are obviously related to the quality of the assumed interaction potential between atoms. The traditional approach to this problem has been to assume a functional form for the interaction potential whose parameters are determined by fitting either a set of experimental data or accurate quantum mechanical calculations. This approach has played a major role in the development of MD, and we expect that it will continue to do so in the future. Effective potentials are, in fact, very simple to use and computationally very effective. This makes them irreplaceable for the study of large systems and/or long-time behavior. Also, they offer a dependable description of the system in many cases. However, when the chemistry of the system is more complex, the effective-potential approach is faced with severe difficulties. In such cases, the potential fitted to a set of properties can be totally inadequate to describe other properties. Similar considerations apply to potentials derived from quantum chemical calculations. These calculations are, in fact, relevant to the gaseous phase where the chemistry is often very different from that in the bulk.

A natural way out is to try and develop a practical scheme to calculate, with a reliable theory, the full many-body interatomic forces during the MD run. Since such a scheme does not involve any empirical assumption, we will call this approach *ab-initio* molecular dynamics. In the present implementation of the method [1], the calculation of the many-body interatomic forces is done within the local density approximation (LDA) to the density functional (DF) theory [2]. This theory provides a convenient compromise between accuracy and computational convenience. Experience has shown that LDA-DF calculations can be very reliable in predicting equilibrium theories, yet they are simpler to implement than other quantum chemical methodologies. However even at the level of the LDA-DF theory, the calculational challenge posed by this approach is formidable and appropriate ways have to be found to make it feasible. Our proposed solution is to map this combined electron-ion problem into an effective classical problem that is simpler to solve. It must be stressed that no additional approximations are introduced by the mapping.

In this note we will review the basic aspects of the method and present some recent applications to the physics of carbon [3] and of some I-IV alloys [4].

2. Ab-Initio Molecular Dynamics and Density Functional Theory

Let us denote by $\Phi(\{\mathbf{R}_I\})$ the Born Oppenheimer (BO) potential energy surface of a multiatomic system, whose spatial configurations are $\{\mathbf{R}_I\}$. The MD trajectories are solutions of Newton's equations of motion:

$$M_I \ddot{\mathbf{R}}_I = -\frac{\partial \Phi}{\partial \mathbf{R}_I} \tag{1}$$

In *ab-initio* MD, $\Phi(\{\mathbf{R}_I\})$ is calculated from the total ground-state energy. The theorem of Hohenberg and Kohn [2], which is the cornerstone of DF theory, ensures that the ground-state energy of an inhomogeneous electron gas is given by the minimum of the total energy functional $E\left[\rho_e(\vec{r}), \{\mathbf{R}_I\}\right]$ relative to the electron charge density $\rho_e(\vec{r})$. It is convenient to express $\rho_e(\vec{r})$ in terms of a set of occupied orthonormal single-particle orbitals $\{\psi_i\}$ as $\rho_e(\vec{r}) = \Sigma_i^{occ}|\psi_i(\vec{r})|^2$. The Euler equations associated with the minimum principle can be written in the Kohn-Sham form [5]:

$$\left(-\frac{1}{2}\nabla^2 + V^{ext}(\vec{r}) + V^H(\vec{r}) + V^{xc}(\vec{r})\right)\psi_i(\vec{r}) = \epsilon_i \psi_i(\vec{r}). \tag{2}$$

$V^H(\vec{r})$ is the Hartree potential and $V^{xc}(\vec{r}) = \delta E^{xc}/\delta\rho_e(\vec{r})$ the exchange and correlation potential. These equations form a set of self-consistent single-particle equations whose terms are exactly known with the exception of $E^{xc}[\rho]$, which contains the nontrivial part of the many-fermion electronic problem. In order to make the problem soluble, the so-called local density approximation is adopted.

In principle, one could think of performing *ab-initio* simulations by executing the following operations at each MD move: (i) Solve Eq. (1) for a given ionic configuration $\{R_I\}$; (ii) compute the forces acting on the ions, and (iii) solve Newton's equations of motion: $M_I \ddot{\mathbf{R}}_I = -\nabla_{R_I}\Phi$. However, for most systems this procedure is computationally too demanding.

We do not follow this approach. Instead [1], we treat $E\left[\rho_e(\vec{r}), \{\mathbf{R}_I\}\right]$ as the potential energy surface of a fictitious classical dynamical system consisting of the nuclear plus the electronic degrees of freedom, i.e. $\{\mathbf{R}_I\}$ and $\{\psi_i\}$, respectively. The corresponding equations of motion are:

$$\mu\ddot{\psi}_i(\mathbf{r},t) = -\frac{1}{2}\frac{\delta E}{\delta\psi_i^*(\mathbf{r},t)} + \sum_j \Lambda_{ij}\psi_j(\mathbf{r},t) \tag{3}$$

$$M_I\ddot{\mathbf{R}}_I = -\frac{\partial E}{\partial\mathbf{R}_I(t)} \tag{4}$$

for electrons and ions, respectively. μ is an adjustable parameter, M_I are nuclear masses, and Λ_{ij} are Lagrangian multipliers used to impose the orthonormality between the wave functions.

In general, the nuclear trajectories obtained in this way and those generated with Eq. (1) do not coincide, but one can choose the parameter μ such that the electronic dynamics is sufficiently faster than the nuclear one. As a consequence, there is little mixing between slow nuclear and fast electronic degrees of freedom. Thermal equilibration requires a very long time, and nuclear trajectories, initially lying on the BO surface, deviate only very slowly from it. This dynamics may be called classical adiabatic dynamics. It allows simulation of the motion of the nuclei without the need for a separate self-consistent electronic minimization at every time step. Metallic or level-crossing systems pose a special challenge and special precautions have to be taken.

A distinctive feature of *ab-initio* MD is that one generates the nuclear trajectories and the corresponding ground-state electronic charge densities and potentials simultaneously. One can therefore follow directly the evolution of the chemical bonds resulting from atomic motion, and compute electronic properties consistently.

3. Selected Applications: Melting of C and KSi

Although the method has been applied to a large variety of systems, we will review here only two very recent applications to the properties of liquid carbon [3] and to molten KSi [4].

The importance of C in many areas of pure and applied science hardly needs to be underlined. Yet many important facts even of elementary aspects of its phase diagram are poorly understood [6]. Here we focus on a very qualitative feature such as the slope of the high-pressure phase boundary between diamond and liquid C. Experiments on this part of the phase diagram are very difficult to perform in view of the extreme range of pressures and temperatures involved. This has led to a long-standing controversy on whether C behaves like Si or Ge, namely, whether its melting temperature decreases with pressure. *Ab-initio* methods can be of invaluable help in this respect since they can be used with confidence in those regions of the phase diagram where experiments are either lacking or difficult. Conventional MD methods would have to overcome severe difficulties, owing to the lack of empirical data and the difficulty of modelling a very complex bonding behavior.

In Fig. 1 we plot the temperature dependence of the pressure at constant volume in the solid and liquid phases. This has been obtained in an

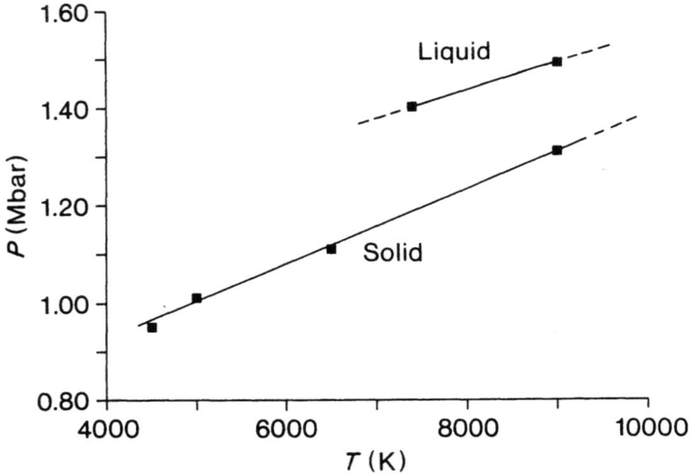

Fig. 1. Carbon pressure as a function of temperature in the solid and liquid phases. From [3]. Copyright 1990 by the AAAS.

ab-initio MD simulation of 54 atoms of C arranged in the diamond structure in the solid phase. The technical details are described in Ref. [3]. In the figure, a positive jump in pressure upon melting is clearly seen. This is related to the fact that the simulation has been carried out at constant volume. In a constant-pressure experiment, this would have led to an increase of the volume upon melting. This is in contrast to the observed behavior for Si and Ge. Application of the Clausius Clapeyron equation then leads to the conclusion that the behavior of C is more standard and that an increase in P leads to an increase in the melting temperature.

Our finding has some important consequences regarding the physics of planets. C is solid in outer planets such as Uranus or Neptune, where it is believed to be at $P \approx 6$ Mbar. Also, any free C in the earth mantel is expected to be solid. Finally — and maybe of more practical relevance — diamond anvil cells have a much larger range of stability than previously assumed.

We shall now briefly consider another system in which binding is rather complex, namely molten KSi. This system belongs to a family of very interesting compounds obtained by alloying alkali metals with group-IV elements. Extensive studies on the alloys formed with the heavier group-IV elements Sn and Pb have demonstrated the existence of several anomalies in the thermodynamic and transport properties. Neutron-scattering experiments have also shown a pronounced first diffraction peak [7,8]. These anomalies have been associated [9] with the existence in the liquid of stable polianionic complexes of the form $(\mathrm{Sn})_4^{4-}$ and $(\mathrm{Pb})_4^{4-}$. In the chemical literature, these are called Zintl ions and are known to be present in the solid phases of these compounds, including KSi.

We have simulated the behavior of KSi with an *ab-initio* MD simulation [4] at $T = 1700$ K, very close to the melting point of Si. The

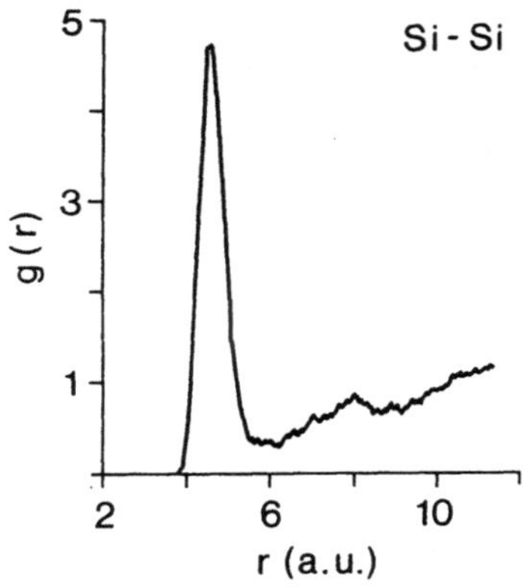

Fig. 2. Silicon-silicon pair correlation function in KSi at 1700 K.

resulting Si-Si partial correlation function is shown in Fig. 2. It clearly shows a very sharp first peak. The coordination number is ≈ 3. This is apparently in agreement with the expected picture that also in the liquid Zintl-like $(Si)_4^{4-}$ ions are formed.

A closer inspection of the system shows, however, that the picture is not so simple and that Zintl-like ions do not exist as stable entities in the liquid. Instead an extended Si network is formed, which is locally tetraedrically coordinated. Clearly, this calls for caution in the interpretation of the experiments.

In conclusion, we have shown with some selected applications that *ab-initio* MD is a very powerful tool and that qualitatively new fields of investigations are now accessible.

References

[1] R. Car and M. Parrinello, Phys. Rev. Lett. **55**, 2471 (1985).

[2] For a review, see, e.g., *Theory of the Inhomogeneous Electron Gas* edited by S. Lundquist and N. H. March (Plenum Press, New York, 1983).

[3] G. Galli, R. M. Martin, R. Car, and M. Parrinello, Science (in press).

[4] G. Galli and M. Parrinello, J. Phys.: Condens. Matter (to appear).

[5] W. Kohn and L. Sham, Phys. Rev. **140**, A1133 (1965).

[6] For a review, see, e.g., F. P. Bundy, Physica A **156**, 169 (1989) and references therein.

[7] B. P. Alblas, W. van der Lugt, J. Dijkstra, W. Geertsma, and C. van Dijk, J. Phys. F: Met. Phys. **13**, 2465 (1983).

[8] M.-L. Saboungi, R. Blomquist, K. J. Volin, and D. L. Price, J. Chem. Phys. **87**, 2278 (1987); H. T. J. Reijers, W. van der Lugt,

and C. van Dijk, Physica B **144**, 404 (1987); H. T. J. Reijers, M.-L. Saboungi, D. L. Price, J. W. Richardson, Jr., K. J. Volin, and W. van der Lugt, Phys. Rev. B **40**, 6018 (1989).

[9] W. Geertsma, J. Dijksta, and W. van der Lugt, J. Phys. F: Met. Phys. **14**, 1833 (1984).

Density-Functional Molecular Dynamics Calculations for Defects in Si and Al

T. Oguchi and T. Sasaki

National Research Institute for Metals, Materials Physics Division, 2-3-12 Nakameguro, Meguro-ku, Tokyo 153, Japan

Abstract. The density-functional molecular dynamics method has been applied to defects in Si and Al, with a supercell model. Prior to the applications, the method and some numerical algorithms involved are briefly described and examined in small-scaled simulations for Si trimer and dimer systems. A preliminary result of finite-temperature simulations for a Si vacancy shows that a local stable atomic configuration around the vacancy has a C_{2v} symmetry, in which two atoms among the nearest neighbors form a pair while two remain apart. In Al, stacking-fault energies are evaluated by taking account of atomic relaxations. The results are discussed by comparing with experiments.

1. Introduction

In molecular dynamics (MD) simulations, the classical trajectories of interacting particles are calculated by solving Newtonian equations to extract their dynamical or statistical properties. Physical quantities in the system are governed by the adiabatic potential, which is usually approximated or modeled by a simple sum of interatomic potentials. It is often very difficult to determine proper potentials within the limited configurational space available.

The local-density functional (LDF) theory provides a fairly good description for the total energy of a system of electrons and nuclei, which gives the adiabatic potential for nuclear motions within the Born-Oppenheimer (BO) approximation. However, the total energy calculation itself is a time-consuming task and in most cases its direct application to an MD simulation is far beyond a tractable limit with the present computer environments.

Recently, Car and Parrinello [1] have proposed a new method which permits a dynamical simulation of atoms on an adiabatic potential surface determined within the framework of the LDF theory. This method, which we call the density-functional molecular dynamics (DF-MD) method, opened a new era in the fields of molecular dynamics and electronic-structure calculations. Many challenging applications have been successfully done on a wide variety of systems and problems such as molecules or clusters [2-5], impurities in solids [6], surfaces or grain boundaries [7-10], liquid or amorphous phases [11-15], ceramics [16], liquid solutions [17], diffusion [18] and thermal expansion [19]. Simplified or related versions of the DF-MD method have been also proposed [3-5, 7, 20-25].

The usage of the DF-MD method may be categorized into the following three types of calculations:

(a) electronic structure calculations for large systems
(b) structural determinations by dynamical simulated annealing (DSA) or steepest descent (SD) techniques
(c) dynamical simulations at finite temperatures.

It is well recognized that dynamical or iterative techniques to solve a Schrödinger-type equation require less operations and storage as a function of the number of basis sets than the diagonalization technique usually used. Then, the efficiency becomes prominent when applied to large (many-atom) systems (type (a)). In a microcluster, surface or impurity system, structural variations or relaxations should be determined in a proper way since they dominate the electronic structure and related properties (type (b)). Applications to study finite-temperature properties might be most interesting in the sense that the adiabatic potential is determined from first principles (type (c)). However, DF-MD simulations for a relatively small number of atoms (~ 100) and for a very short time (~ picosecond) have only been possible so far.

In this paper, we apply the DF-MD method to defects in Si and Al and investigate the stable local structure around the defects and some dynamical properties. Before the applications, much simpler molecule systems, a Si trimer and dimer, are chosen as test simulations, in which basic aspects and physical backgrounds of the method are studied in detail. Numerical parameters and algorithms used in the simulations are also examined.

As for a defect system in Si, we present a preliminary result for a vacancy. Atomic configurations around the vacancy site are determined based on the simulations at room temperature. A phonon property is also discussed.

The formation energies of twin and intrinsic stacking faults in Al are evaluated by taking lattice relaxations into account. The results are in good agreement with experiment.

2. The Density-Functional Molecular Dynamics Method

In the DF-MD method, atomic and electronic degrees of freedom are treated on a common footing. The LDF total energy can be expressed as $E[\{\psi_i\},\{\mathbf{R}_n\}]$, where ψ_i is the electronic wave function for the i-th state and $\mathbf{R}_n$ is the coordinate of the n-th atom. In a rigorous LDF sense, the exact BO surface should be determined by a variational procedure of $E[\{\psi_i\},\{\mathbf{R}_n\}]$ with respect to ψ_i. However, where a global minimization of $E[\{\psi_i\},\{\mathbf{R}_n\}]$ is concerned, a path to the equilibrium point is irrelevant. Furthermore, as long as a deviation from the exact BO surface is negligible compared with the kinetic energy of atoms (temperature in thermodynamics), a proper dynamics might be realized on the potential surface, $E[\{\psi_i\},\{\mathbf{R}_n\}]$.

Corresponding to the potential surface spanned by $\{\psi_i\}$ and $\{\mathbf{R}_n\}$, we may set up the following Lagrangian:

$$L = K_e + K_a - E[\{\psi_i\},\{\mathbf{R}_n\}], \tag{1}$$

$$K_e = \mu \sum_i \int d\mathbf{r} \; |\dot{\psi}_i|^2, \tag{2}$$

$$K_a = \frac{1}{2} \sum_n M_n \; |\dot{\mathbf{R}}_n|^2, \tag{3}$$

where the dot means time derivative and μ and M_n are the masses of the electron and the n-th atom with respect to their classical motions. The wave functions are subject to an orthonormal constraint,

$$\int d\mathbf{r} \;\; \psi_i^* \, \psi_j = \delta_{ij}. \tag{4}$$

Consequently, the equations of motion are written as

$$\mu \ddot{\psi}_i = -\frac{\delta E}{\delta \psi_i^*} + \sum_i \varepsilon_{ij}\, \psi_j, \tag{5}$$

$$M_n\, \ddot{\mathbf{R}}_n = -\nabla_n E. \tag{6}$$

ε_{ij} is a Lagrange multiplier to hold the orthonormality (eq. (4)). Note that the sum of two kinetic and LDF total-energy terms $E_G = K_a + K_e + E$, which we call here the grand total energy, is a conserved quantity of the motion.

Equations (5) and (6) can be solved by using a standard MD algorithm such as Verlet's method [26]. In this study, we adopt a plane-wave expansion for the wave functions and norm-conserving pseudopotentials given by Bachelet *et al.* [27]. With use of the plane-wave basis set, forces acting on each atom are simply given by the same expression as the Hellmann-Feynman force *without* corrections, since ψ_i and $\mathbf{R}_n$ are independent variables of the motion. It should be noted that the way to evaluate ε_{ij} is one of the important key points for evaluating accurate dynamics. It is found that the iterative scheme [28] is more appropriate for the purpose than the Gram-Schmidt scheme.

When an electronic-structure calculation or a global minimization rather than dynamics is concerned, we may use first-order differential equations instead of Newtonian ones [21-24]:

$$\dot{\psi}_i = -\frac{\delta E}{\delta \psi_i^*} + \sum_i \varepsilon_{ij}\, \psi_j. \tag{7}$$

Equation (7) is solved by a so-called steepest descent (SD) or conjugate-gradient technique, for example. In this case, how to hold the orthonormal condition is irrelevant and the Gram-Schmidt scheme sometimes provides a better convergence because of its symmetry-breaking nature [29]. If the SD-type minimization procedure is used for atomic degrees of freedom, care must be taken to seek the absolute minimum because there may be many minima in the potential surface in most cases.

3. Test Calculations for a Si Trimer and Dimer

In order to investigate basic problems in the DF-MD method, we apply it to simple molecular systems. In particular, a numerical accuracy and stability and a choice of appropriate parameters and algorithms involved are examined.

The first case is a Si trimer, which has been studied with a highly-precise quantum-chemical approach [30]. As an initial atomic configuration we consider a triangle form with the length of two bonds $d = 2.35$ Å and the angle between these bonds $\theta = 90°$. A trimer molecule is placed in a supercell with a sufficiently large volume. Starting with randomly-distributed coefficients of the basis functions, the BO surface can be easily reached for this initial configuration by using either SD or DSA. Since the assumed atomic configuration is not stable, each atom starts to move internally once atomic degrees of freedom are allowed. The total and kinetic energies for a 5000-step run are plotted in Fig. 1. In this simulation, the fictitious mass μ = 200 a.u., the time step Δt = 4 a.u. and the energy cut-off of the plane waves ε_c = 6 Ry are used. (Throughout the paper, Rydberg atomic units are used unless otherwise noted.) During the whole simulation the grand total energy E_G is conserved within seven decimal digits. As seen in Fig. 1, an oscillatory behavior of K_e indicates that the wave functions follow accurately the atomic motion. An amplitude of K_e much smaller than K_a implies that the dynamics stays closely on the BO surface. It can be clearly recognized in Fig. 1 that the atomic oscillation contains two modes with relatively fast and slow frequencies. In order to clarify

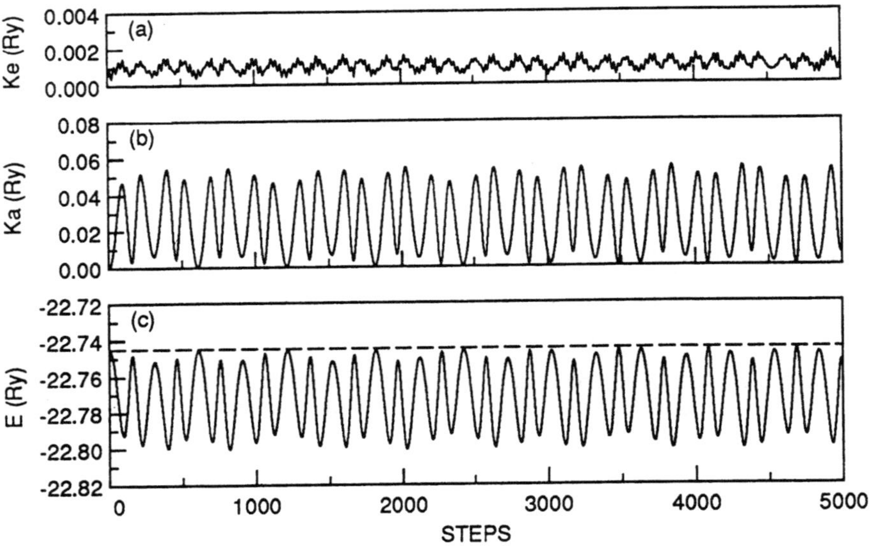

Fig. 1 Time dependence of the kinetic and LDF total energies in a Si trimer system: (a) K_e, (b) K_a and (c) E (solid line) and $E_G = K_a + K_e + E$ (broken line). Note a scale difference in (a) and (b).

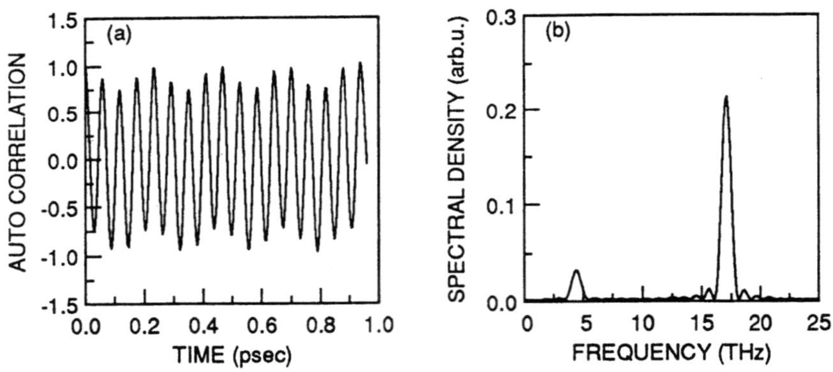

Fig. 2 (a) Velocity auto-correlation function and (b) its spectral density in a Si trimer. Peaks at 4.4 THz and 17.1 THz in (b) correspond to bending and breathing modes, respectively.

these modes, the velocity auto-correlation function and its spectral density are calculated and shown in Fig. 2. The spectral density has two peaks at 4.4 THz and 17.1 THz, which correspond to bending and breathing motions, respectively. It is quite interesting that these two modes are already very close to the acoustic and optical frequencies in bulk Si. With use of the SD technique described in § 2, the most stable configuration can be determined, by starting from several initial ones. The equilibrium structural parameters determined are d = 2.14 Å and θ = 80.6°,

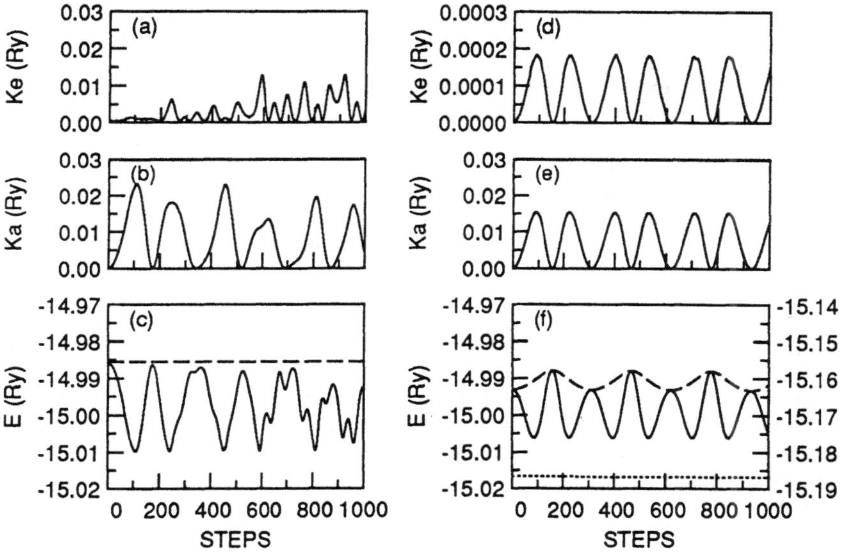

Fig. 3 Time dependence of the kinetic and LDF total energies in a Si dimer system without ((a)-(c)) and with ((d)-(f)) Fermi broadenings: (a) K_e, (b) K_a, (c) E (solid line) and $E_G = K_a + K_e + E$ (broken line), (d) K_e, (e) K_a and (f) E (solid line), $E_G = K_a + K_e + E$ (broken line) and $K_a + K_e + F$ (dotted line).

which are in good agreement with the result of an *ab initio* molecular-orbital calculation ($d = 2.17$ Å and $\theta = 77.8°$) [30].

Unlike the case of a trimer, a Si dimer is difficult to treat with the DF-MD method. Figure 3(a)-(c) shows a time dependence of the total and kinetic energies in a simulation for a Si dimer system. After 200 steps, K_e becomes larger up to the same order as K_a and the atoms move in a wrong way, though E_G is well conserved. This is because only integer occupations are allowed in the simulation. It is known that the highest occupied state is a π state with a partial filling near the equilibrium atomic distance, and a level crossing with the other electronic state happens to occur at smaller distances [4]. As already pointed out in recent work [15, 22, 31], it is necessary to introduce non-integer occupations to prevent such numerical instability in this situation as well as a metallic system. By allowing non-integer occupations with use of the Fermi function, a much better dynamics can be achieved as shown in Fig. 3(d)-(f). In this case, all the kinetic energies plus a free energy, which contains a configurational entropy of electrons, should be conserved, instead of E_G. Note the difference of the energy scale in K_a and K_e in Fig. 3. Non-integer occupations are considered in the case of Al in § 5.

4. An Application to a Vacancy in Bulk Si

It has been accepted for a vacancy in Si from electron-paramagnetic-resonance and deep-level-transient-spectroscopy measurements [32] that there exist several charged states and symmetry-lowering lattice distortions, depending on the position of the chemical potential. On the other hand, electronic structure calculations have been done extensively for the vacancy with the use of a Green function method [33]. Among them lattice relaxations of the neighboring atoms have been incorporated within a Keating model.

Since an ideal, unrelaxed vacancy has a T_d symmetry, dangling bonds of the nearest neighbor (NN) atoms surrounding the vacancy would form a_1 and t_2 states. The former is resonating inside the valence band while the latter is partially occupied with two electrons in a neutral case, being a deep impurity level. Such a partial occupation may induce a Jahn-Teller (JT) lattice distortion to a lower symmetry accompanied by an energy splitting of the t_2 state. It is understood that a distortion to tetragonal D_{2d} symmetry, in which double pairs are formed with the four NN atoms, may occur in the case of a neutral vacancy. A coupling of the occupation to the JT lattice distortion is well known as a negative-U problem.

We set up a 32-atom supercell with the diamond structure as a host and remove an atom from it to model a vacancy system. We use Nosé's algorithm [34] to maintain the system at a given temperature, T, for finite-temperature simulations. A degree of freedom allowing an energy transfer between the system and a heat reservoir is added to the equations of motion. All simulations have been done with $T = 300$ K and $\Delta t = 2$ a.u.

For the ideal configuration, forces acting on the NN atoms are negligibly small, since only very small JT forces exist due to a large distance between the NN atoms (~ 3.84 Å) and two opposite breathing forces originating from the electrostatic and electronic terms are almost canceled out. However, the NN atoms have more room to move because of the existence of the vacancy. Actually, a simulation at room temperature shows large movements of the NN atoms. Figure 4 plots the neighboring distances measured from one of the NN atoms and a reduction of the distance clearly seen indicates the formation of a loosely-bound pair with the other NN. It is interesting that the minimum distance of the pair reaches almost to the ideal NN bond distance (~ 2.35 Å). But no pair formation has been observed for the two remaining NN atoms during the present simulation (~ 1.2 psec). Instead, they move slightly further apart. Consequently, the stable local configuration may have a C_{2v} symmetry rather than a tetragonal D_{2d}. A relatively high energy barrier for making double pairs within D_{2d} has been found by a static total-energy calculation for the same supercell, where atomic positions of the NN

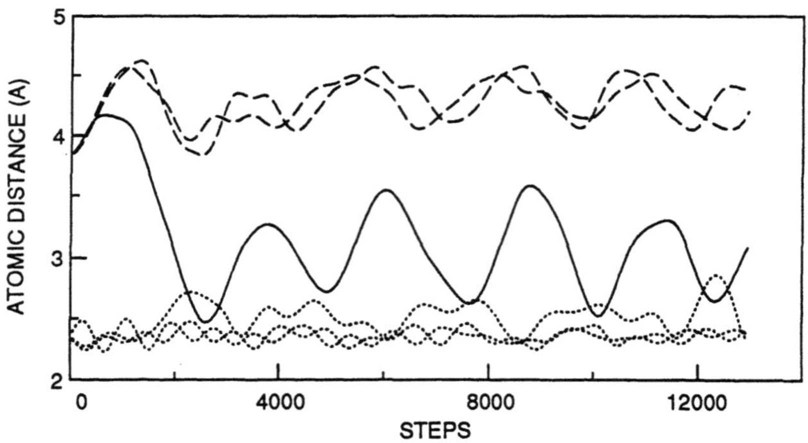

Fig. 4 Atomic distances measured from one of the nearest neighbor atoms surrounding the vacancy in Si. A large reduction of the distance to another nearest neighbor means a pairing formation (solid line) while the distances to the two remaining nearest neighbors become slightly larger than the ideal ones (broken lines). The bond distances between the nearest and three second nearest neighbors are also shown with dotted lines.

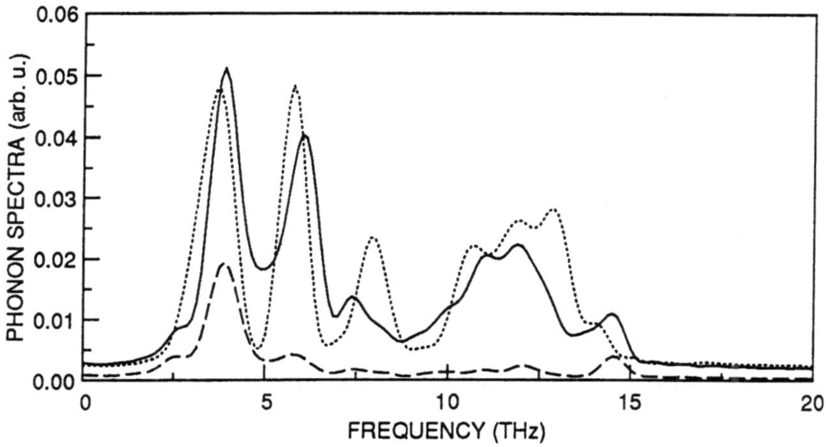

Fig. 5 Phonon spectra of bulk Si with (solid line) and without (dotted line) a vacancy. A projected spectrum on the nearest neighbor sites surrounding the vacancy is drawn with a broken line.

atoms are given and the others are frozen at the ideal sites. Further investigations with the use of a larger supercell and a higher energy-cutoff for the plane-wave basis set should be necessary to make a conclusive statement for the stable ground-state configuration.

A phonon density of states (DOS) was calculated from the velocity autocorrelation function for the host and vacancy systems, as drawn in Fig. 5 along with a projected phonon DOS on the NN sites. The phonon spectra show noticeable differences between the two systems. A shoulder of the projected density at 2.5 THz is due to the slow movement of the NN atoms as already discussed above. More interestingly, another structure can be seen at the high frequency region around 15 THz, where a small enhancement from the host may be attributed to an optical, breathing mode hardening of bonds between the NN and second-nearest neighbor atoms due to a charge density redistribution into these bonds.

5. An Application to Stacking Faults in Al

To study structural and mechanical properties in Al, many *empirical* pseudopotential calculations within the perturbation theory have been performed and qualitative understanding of the properties has been obtained as well in other simple metals [35-37]. Some studies were intended to evaluate energies associated with several types of defects in Al [38]. Although the perturbation theory gives almost the right order of the defect energy in most cases, the results strongly depend on how the defect is modeled and how the energy is evaluated even if one uses the same pair potentials. Therefore, it is still an important and challenging subject to study properties of defects in Al in a *non-empirical* way.

In this section, defects such as twin and intrinsic stacking faults, TSF and ISF, are studied with the DF-MD method. The supercell geometry is again used for modeling the fault systems. The fault energies are evaluated from total-energy differences between the perfect fcc host and the TSF and ISF systems by taking account of atomic relaxations and comparing with experimental and other theoretical results. Since the atomic relaxations involved in the stacking faults are

	model I	model II
fcc	...*ABCABC*...	...*ABCABCABCABC*...
hcp	...*ABABAB*...	...*ABABABABABAB*...
TSF	...*ABCACB*...	...*ABCABCACBACB*...
ISF	...*ABCBC*...	...*ABCABCBCABC*...

Fig. 6 Sequences of the (111) plane stackings of fcc, hcp, twin (TSF) and intrinsic (ISF) stacking faults in the supercell models.

expected to be small in a simple metal such as Al, the SD scheme was used to determine the equilibrium configuration instead of DSA.

Figure 6 shows a summary of supercell models used for TSF and ISF as well as host fcc. Assuming the three different (111) close-packed planes of fcc as *A*, *B* and *C*, supercells (model I) for TSF and ISF are represented as (111) sequences of ...*ABCACB*... and ...*ABCBC*..., respectively, while the perfect fcc (hcp) lattice is ...*ABCABC*... (...*ABABAB*...). Double-size supercells (model II) are also used for checking the convergency of model size.

Since Al has rather complicated Fermi surfaces and the total-energy difference to be discussed is so small, it is important to test the convergence of *k*-point sampling in the Brillouin zone before calculating the fault energies. For this purpose, we have calculated the structure energy of hcp defined as a total-energy difference between hcp and fcc using uniform mesh points and special points [39]. The c/a ratio of the lattice constants was taken to be an ideal value and the model-I supercell (six Al atoms) was used. The calculated structure energies with several choices of *k*-points are listed in Table I. It is concluded that the twelve special point scheme can be used to discuss the total-energy difference for the present systems within an error of a few mRy. All the results presented below are done with the twelve special points.

Table I Total-energy difference between hcp and fcc phases in Al as changing number of *k* points (N_k) in the full Brillouin zone. (mRy/atom)

	uniform mesh		special point
N_k	16	32	12
ΔE	3.8	3.0	4.6

It is found that, for the ideal, unrelaxed TSF and ISF systems, forces acting on atoms are quite small, being at most a few mRy/Bohr near the faults and rapidly decreasing with distance from the faults. The direction of the forces is in such a way that the two *B* or *C* planes existing on both sides of *A* get further apart in TSF while the two planes nearest to the faults become closer in ISF. Anyhow, the obtained relaxation near the faults is only order of 0.01Å, resulting in a change of a few erg/cm^2 in the total energy. Related to this very small relaxation, the present two supercell models show essentially equivalent results for the forces and energies.

Table II Twin (TB) and intrinsic (ISF) stacking-fault energies of Al. (erg/cm^2)

	TB	ISF
present	110	230
LMTO+	135	250
perturbation*	8-100	31-195
experiment at 300K*	75-100	135
extrapolated to 0K*	100-125	160

*: reference 38
+: reference 40

The fault energies evaluated from the total-energy difference for TSF and ISF in Al are summarized in Table II together with results of total-energy calculations by the LMTO method with a supercell [40] and by the perturbation theory based on the empirical pseudopotential, and experimental data [38]. The calculated value for TSF is in fairly good agreement with experiment and another first-principles calculation with LMTO. The ISF energy seems to be overestimated, though the same trend has been obtained by the LMTO calculation. As already noted, the atomic relaxations in both TSF and ISF are quite small. This result can be interpreted by the fact that these faults maintain the close-packed structure and charge redistributions are very small because of the screening effect.

It is a rather surprising result that the perturbation theory gives relatively fair results for the stacking-fault energy. This may be only true for the faults in which the perturbation introduced is small and this is not the case for a vacancy or grain-boundary, for instance. As claimed previously [38], the total energy calculated for defects by the second-order perturbation theory is very sensitive to how it is evaluated because of the long-range nature of the pair potential.

Acknowledgements

The authors would like to thank S. Blügel, K. Terakura and H. Katayama-Yoshida for invaluable discussions.

References

[1] R. Car and M. Parrinello, Phys. Rev. Lett. **55** 2471 (1985).
[2] D. Hohl, R.O. Jones, R. Car and M. Parrinello, J. Chem. Phys. **89** 6823 (1988).
[3] F.S. Khan and J.Q. Broughton, Phys. Rev. B**39** 3688 (1989).
[4] O.F. Sankey and D.J. Niklewski, Phys. Rev. B**40** 3979 (1989).
[5] J. Broughton and F. Khan, Phys. Rev. B**40** 12098 (1989).
[6] T. Sasaki, T. Oguchi and H. Katayama-Yoshida, Phys. Rev. B (to be published).

[7] O.F. Sankey and R.E. Allen, Phys. Rev. B33 7164 (1986).
[8] M. Needels, M.C. Payne and J.D. Joannopoulos, Phys. Rev. Lett. **58** 1765 (1987).
[9] M. Needels, M.C. Payne and J.D. Joannopoulos, Phys. Rev. B**38** 5543 (1988).
[10] M.C. Payne, P.D. Bristowe and J.D. Joannopoulos, Phys. Rev. Lett. **58** 1348 (1987).
[11] R. Car and M. Parrinello, Phys. Rev. Lett. **60** 204 (1988).
[12] G. Galli, R.M. Martin, R. Car and M. Parrinello, Phys. Rev. Lett. **62** 555 (1989).
[13] G. Galli, R.M. Martin, R. Car and M. Parrinello, Phys. Rev. Lett. **63** 988 (1989).
[14] I. Stich, R. Car and M. Parrinello, Phys. Rev. Lett. **63** 2240 (1989).
[15] G.X. Qian, M. Weinert, G.W. Fernando and J.W. Davenport, Phys. Rev. Lett. **64** 1146 (1990).
[16] D.C. Allan and M.P. Teter, Phys. Rev. Lett. **59** 1136 (1987).
[17] A. Selloni, P. Carevali, R. Car and M. Parrinello, Phys. Rev. Lett. **59** 823 (1987).
[18] F. Buda, G.L. Chiarotti, R. Car and M. Parrinello, Phys. Rev. Lett. **63** 294 (1989).
[19] F. Buda, R. Car and M. Parrinello, Phys. Rev. B**41** 1680 (1990).
[20] M.R. Pederson, B.M. Klein and J.Q. Broughton, Phys. Rev. B**38** 3825 (1988).
[21] J.L. Martins and M.L. Cohen, Phys. Rev. B**37** 6134 (1988).
[22] C. Woodward, B.I. Min, R. Benedek and J. Garner, Phys. Rev. B**39** 4853 (1989).
[23] I. Stich, R. Car, M. Parrinello and S. Baroni, Phys. Rev. B**39** 4997 (1989).
[24] M.P. Teter, M.C. Payne and D.C. Allan, Phys. Rev. B**40** 12255 (1989).
[25] D. Hohl and J. Harris, Bull. Am. Phys. Soc. **35** 709 (1990).
[26] L. Verlet, Phys. Rev. **159** 98 (1967).
[27] G.B. Bachelet, D.R. Hamann and M. Schlüter, Phys. Rev. B**26** 4199 (1982).
[28] J.P. Ryckaert, G. Ciccotti and H.J.C. Berendsen, J. Comput. Phys. **23** 327 (1977).
[29] M.C. Payne, M. Needels and J.D. Joannopoulos, Phys. Rev. B**37** 8138 (1988).
[30] K. Raghavachari and V. Logovinsky, Phys. Rev. Lett. **55** 2853 (1985).
[31] G.W Fernando, G.X. Qian, M. Weinert and J.W Davenport, Phys. Rev. B**40** 7985 (1989).
[32] *For a review of the experimental works on the Si vacancy, see for example*: G.D. Watkins, J.R. Troxell and A.P. Chatterjee, Inst. Phys. Conf. Ser. No. 46: Chap. 1, 16 (1979).
[33] *For a review of the band-theoretical work on the Si vacancy, see for example*: J. Bernholc, N.O. Lipari, S.T. Pantelides and M. Scheffler, Inst. Phys. Conf. Ser. No. 59, 1 (1981); G.A. Baraff, E.O. Kane and M. Schlüter, *ibid.*, 19 (1981).
[34] S. Nosé, J. Chem. Phys. **81** 511 (1984); S. Nosé, Molec. Phys. **52** 255 (1984).
[35] W.A. Harrison, Phys. Rev. **136** A1107 (1964).
[36] R.W Shaw, J. Phys. C: Solid St. Phys. **2** 2335 (1969).
[37] M.A. Coulthard, J. Phys. C: Solid St. Phys. **3** 820 (1970).
[38] J.P. Simon, J. Phys. F: Metal Phys. **9** 425 (1979).
[39] D.J. Chadi and M.L. Cohen, Phys. Rev. B**8** 5747 (1973).
[40] J.H. Xu, W. Lin and A.J. Freeman, Phys. Rev. B (to be published).

Optimum Adsorption Sites and Electronic Structure of Alkali-Adsorbed Si(001) Surfaces

K. Kobayashi, K. Terakura, and S. Blügel

Institute for Solid State Physics, University of Tokyo, Minato-ku, Tokyo 106, Japan

The first-principles molecular dynamics method was applied to study the atomic and electronic structures of the Na-adsorbed Si(001) surfaces. The coverages studied were $\theta = 1, 1/2$ and $1/6$, with $\theta = 1$ corresponding to the case for one adatom per one surface Si atom.

1. Introduction

The structural determination is an indispensable first step towards understanding material properties from a microscopic view point. For perfect crystals, very precise structural information can be obtained by x-ray diffraction and neutron diffraction, for example. However, the structures associated with defects are generally very hard to determine experimentally. It is particulary so for surfaces and interfaces. Therefore, it is highly desirable that theories provide some useful complementary information in those cases. It turns out from many examples that the electronic structure calculation based on the local density approximation (LDA) in the density functional theory [1] can serve as a potentially very powerful tool for such a purpose. Practically, however, it required a huge amount of computer resource and labor to perform a structural optimization for systems with more than three degrees of freedom.

Substantial progress was made by Car and Parrinello [2] who introduced a new idea of combining the LDA band-structure calculation with the molecular dynamics. We call the method first-principles molecular dynamics (FPMD). Molecular dynamics has been used for structural optimization with a given inter-atomic potential. Nice aspects in FPMD are that the force acting on each atom is calculated quantum mechanically and that not only the atomic structure but also the electronic structure are determined simultaneously. In the present work, we apply FPMD to alkali-adsorbed Si(001) surfaces.

The alkali-adsorbed Si(001) surfaces have been extensively studied both experimentally [3-15] and theoretically [16-22] and yet some fundamental aspects are controversial. Even the saturation coverage of Na or K has not been well defined in the absolute scale: controversies are still going on about the single-layer model (like the Levine model)[3,4,9,10,15] versus the double-layer model [5,6,14]. Stable adsorption sites have not yet been determined. The atomic structure for the 2×3 (or 3×2) ordering [10] is not known. It is also true, however, that new information is actively supplied by experiment and theory and there is a hope that a consistent overall understanding will be achieved sooner or later.

In the following, we first study the energetics of the symmetric dimer and the asymmetric dimer on the clean Si(001)-2×1 surface. The present calculation suggests that the asymmetric dimer is definitely more stable than the symmetric dimer. We then study the stable adsorption sites for adsorbed Na for the coverage $\theta = 1, 1/2$ and $1/6$ by allowing full relaxation of the substrate. The coverage θ is defined such that the full coverage ($\theta = 1$) corresponds to one Na (or K) atom per one surface Si atom. The main results are the following. (Note that all the results are for the zero temperature.)

1)Even at the full coverage ($\theta = 1$), Na can be chemisorbed and the most stable configuration for the adsorbed Na is the combination of HH (the hollow site along the hill of the dimer row) and T3 (the site along the trough on top of the Si atom in the third layer). The surface Si dimer becomes symetric and the dimer bond distance is elongated significantly. A top view of these possible sites is shown in Fig.1.

2)At $\theta = 1/2$, the linear chain configuration with Na at T4 (along the trough on top of the Si atom in the fourth layer) is more stable than the Levine model (Na at HH). However, the energy difference is only 12 meV per dimer.

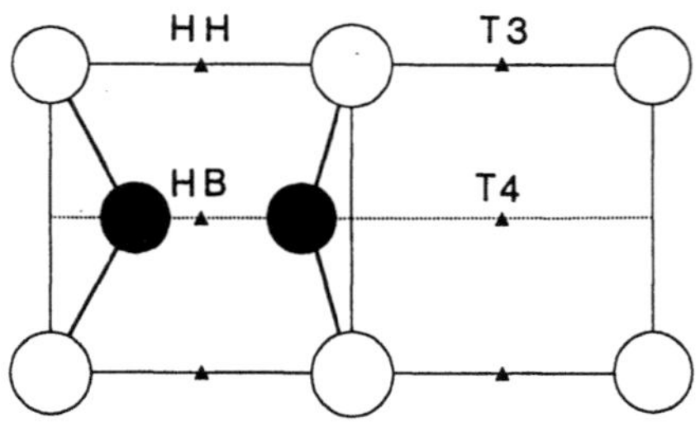

Figure 1. A top view of the asymmetric dimer configuration of the Si(001)-2×1 surface. The filled and the open circles denote the Si atoms in the top-most surface layer and the second layer, respectively. The possible adsorption sites for alkali atoms are denoted as HB, HH, T3 and T4.

3)At $\theta = 1/6$, our preliminary calculation (only 5.0 Ry for the cutoff energy for the plane-wave basis functions) predicts that Na is adsorbed at the bridge site formed by the raised Si atoms of the neighboring two dimers. This is consistent with the scanning tunneling microscopy (STM) study for the K adsorption in a low coverage regime [12,13]. However, our surface unit cell is not large enough to reproduce the observed arrangement of the asymmetric Si-Si dimers around the adsorbed K.

In all of the above calculations, we noticed a rather inconvenient aspect in the theoretically optimized structures: the interatomic distances in the surface region are significantly shorter than those estimated by experiments. A possible origin for this will be discussed.

The present article is organized in the following way. In Sec.2, some details of the method of calculation will be described. Results and discussion will be given in Sec.3 and we conclude in Sec.4.

2. Method of Calculation

The present calculation is based on the LDA in the density functional theory [1] and the norm-conserving pseudopotentiall [23] with a modification by Kleinman and Bylander [24]. We adopt a repeating slab geometry and the plane-wave basis. The substrate is represented by a ten-layer Si(001) slab and a vacuum region of the same thickness is inserted in between Si slabs. The alkali atoms are introduced on each surface of the Si slab. The cutoff energy E_{cut} of the plane-wave basis and the number of k points N_k in the surface Brillouin zone for the k-space integration will be described in each case in the next section.

The optimization of the electronic and the ionic degrees of freedom is performed by a modified version of FPMD proposed by Williams and Soler [25]. We adopt a steepest descent type of algorithm for the electronic degrees of freedom,

$$\gamma\dot{\psi} = -(H - \lambda)\psi, \qquad (1)$$

and use a technique similar to that of Payne et al.[26] for solving eq.(1). Furthermore, a simple mixing of the charge density with equal weight for the input and output is done after each time step in order to suppress oscillations in the process of iteration. As for the ionic motion, we use a standard molecular dynamics scheme. The time step is 20 a.u.($=4.837\times10^{-16}$sec) and the coefficient γ in eq.(1) is 100 in atomic units. (The Hartree atomic units, $\hbar = m_e = e^2 = 1$, are adopted in this paper.) If we adopt eq.(1), the total energy of the ionic system,

$$\frac{1}{2}\sum_n M_n(\frac{d}{dt}R_n)^2 + E[\{\psi\},\{R_n\}], \tag{2}$$

disspates automatically in proportion to γ. The first term of eq.(2) is the kinetic energy of ions and the second term is the sum of the energy of the electronic system in the density functional formalism and the electrostatic energy of the ions.

3. Results and Discussion

3.1 Clean Surface ($\theta = 0$)

Some LEED experiments demonstrate that the clean surface of Si(001) exhibits a c(4×2) pattern which is generally believed to be an ordering state of the asymmetric dimers. On the other hand, the asymmetric dimers can be clearly observed only near the step or some other vacancies by the STM measurements [27,28]. Therefore, it is not absolutely clear whether the asymmetric dimer can be stabilized or not on the clean Si(001) surface. We studied this problem with three different conditions for the cutoff energy and the k-space sampling and show the results in Table 1. Although the energy difference has not converged yet with respect to the cutoff energy E_{cut}, the asymmetric dimer has a lower energy than the symmetric one for all three cases. The energy separation between the π and π^* band gap states becomes wider by about 0.01 a.u. (0.27eV) when the symmetric surface Si-Si dimer is transformed to the asymmetric one. The Si-Si dimer distance in our calculation is 2.30Å being shorter than the experimental estimate [15] of 2.45Å by 6%.

Table 1. The energy differnce between the symmetric dimer model and the asymmetric one for the clean Si(001)-2×1 surface. E_{cut} is the cutoff energy for the plane-wave basis, N_k is the number of k points in the whole surface Brillouin zone and $\Delta E = E(\text{symmetric}) - E(\text{asymmetric})$.

	E_{cut}(in Ry)	N_k	ΔE(eV/dimer)
1	6.25	8	0.068
2	6.25	32	0.053
3	9.00	32	0.161

3.2 Na-Adsorbed Sruface

$\theta = 1$

As this case was discussed in the previous publication [22], we present the calculated results briefly. The computational condition is such that $E_{cut} = 6.25$ Ry and $N_k = 8$. As the system has an insulating gap for $\theta = 1$, this rather small N_k may be sufficient. We found two stable configurations, the combinations HH-T3 and HH-T4, between which the former combination is lower in energy by 28 meV per dimer than the latter . The bridge site of the Si-Si dimer is not a stable position. For both the HH-T3 and HH-T4 cases, a substantial relaxation of the Si substrate is induced (Fig.2): the Si-Si dimers are now symmetric and the dimer distance is 2.65Å (2.51Å) for the HH-T3 (HH-T4) case. The disappearance of the asymmetry is a natural consequence of the full occupancy of the π^* band-gap states, because the energy gain caused by the lowering of the π band is cancelled out by the energy loss caused by the counter movement of the π^* band.

As was mentioned in the Introduction, some experimentalists claim that the saturation coverage will not be $\theta = 1$ but $\theta = 1/2$ for the K and Na adsorption [4,8-10]. These experiments were carried out at room temperatures or even higher, whereas the present calculation is for zero Kelvin. Therefore, a theoretical study on the effect of the finite temperature is needed to give a definitive conclusion about the saturation

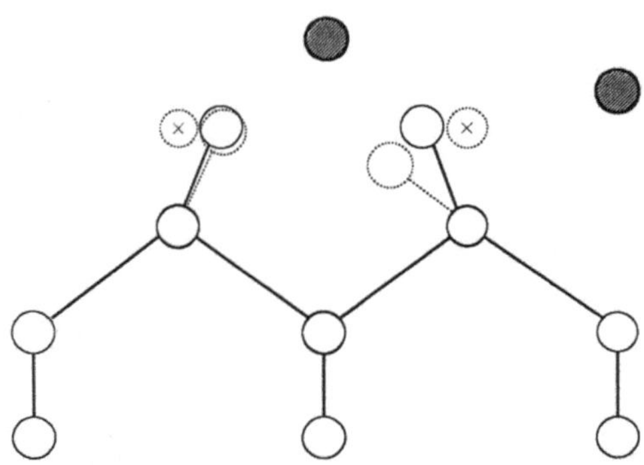

Figure 2. A side view of the optimized atomic configuration for the Na adsorption with the full coverage ($\theta = 1$) in the HH-T3 configuration. The hatched circles represent Na and the others are the substrate Si. The dotted circles with and without a cross denote the ideal bulk terminated Si surface and the asymmetric dimer of the clean Si surface, respectively.

coverage. However, we note that the shortest Na-Si distance in the HH-T3 configuration is 2.62Å (2.81Å) for Na at the HH (T3) site. This relatively short Na-Si distance may imply that Na atoms are chemisorbed rather than physisorbed even for $\theta = 1$.

$\theta = 1/2$

$E_{cut} = 9.0$ Ry and $N_k = 32$ in this case. As the π^* band is half-filled for $\theta = 1/2$, the system is metallic and large N_k is required. The fundamental question in the $\theta = 1/2$ case is whether the alkali atoms are adsorbed along the hill of the dimer chain or along the trough between the dimer chains. For Na adsorption, the pseudopotential calculation by Batra predicted that the HH site is more stable than T4 (T3) site by 0.12 eV (0.22 eV) [21]. On the other hand, a cluster calculation by Ling et al. for K adsorption predicted that the T4 site is most favored [18]. In these calculations, however, the possibility of asymmetry in the Si-Si dimer was not taken into account, whereas it was automatically considered in the present calculation because the starting atomic configuration for the substrate was the one optimized for the clean surface. As is shown in Fig.3, the asymmetry of the Si-Si dimer remains for $\theta = 1/2$ in both cases of the adsorption along the hill and the trough of the dimer rows. There is still an energy gain caused by the asymmetry-induced increase of the $\pi - \pi^*$ band separation as the π^* band is only half-filled. Even if the Na atoms are put fairly close to the T3 sites, they move towards the T4 sites and stop there. Along the hill of the dimer, the HH site is stable

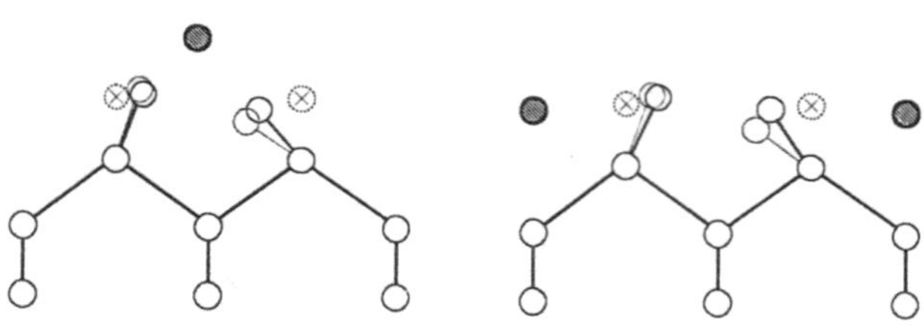

(a) (b)

Figure 3. Side views of the optimized atomic cofiguration for the Na adsorption with $\theta = 1/2$. The hatched circles denote the Na atoms and the others are the substrate Si atoms. The dotted circles with and without a cross denote the ideal bulk terminated Si surface and the asymmetric dimer of the clean Si surface, respectively. (a) is for Na at the HH site and (b) for Na at the T4 site. In the present calculation, case (b) is more stable than case (a) by 12meV/dimer.

while the HB site is unstable. By the present calculation, the T4 adsorption is lower than the HH adsorption by 12 meV/dimer. However, this energy difference is rather close to the limit of the numerical accuracy. Analyses of experiments suggest that the preferential adsorption at T4 rather than HH may not be plausible. We have shown in the previous paper [20] that the T4 site adsorption produces a much smaller lowering of the work function than observed. A recent LEED analysis also concluded that the HH-site adsorption is more favorable than the T4-site adsorption [15]. However, caution may be necessary with this LEED analysis. It was stated in the paper [15] that the double-layer model of the HH-T3 configuration is equally acceptable simply from the LEED analysis. Combining all of them including our result, we speculate that the coexistence of the HH-site adsorption and the T4-(or T3-) site adsorption may be realized. The adatoms along the trough are experimentally rather inactive with regard to the work function, LEED and also STM.

$\theta = 1/6$

The study of this coverage has two aspects. First, we attempt to analyze the low coverage STM image [12,13], which suggests that the adsorption sites for isolated K and Li are none of those sites considered so far. For K, it is the bridge site between the two raised Si atoms of the neighboring Si-Si dimers and for Li, it is the on-top site of the raised Si atom of an asymmetric Si-Si dimer. It is also interesting to note that the adsorbed alkalis seem to enhance the contrast of the STM image associated with the asymmetry of the neighboring Si-Si dimers [12,13]. Second, as the surface unit cell for this coverage is 2×3, the analysis of this case may give a hint to the observed 2×3 structure [10].

Because of the large system size, the present calculation uses a relatively small E_{cut} of 5.0 Ry and yet the number of plane waves is 3072. $N_k = 4$ is also rather small. The present calculation is therefore a preliminary one. The starting atomic configuration for the substrate is the 2×1 -asymmetric dimer structure which was optimized for the clean surface. The Na atom was put slightly off the HB site. The cross in Fig.4a denotes the initial Na position. As the optimization process proceeds, the Na atom moves towards the inter-dimer bridge site as denoted by a hatched sphere in Fig.4a. As mentioned above, this is the position that the STM image [12,13] suggests as the adsorption site for K in the low coverage regime. Figure 4b is a side view containing Na and the Si atoms designated as A, A', B and B' in Fig.4a.

The present coverage $\theta = 1/6$ may not be low enough to make a direct comparison with the experiment and because of that the observed

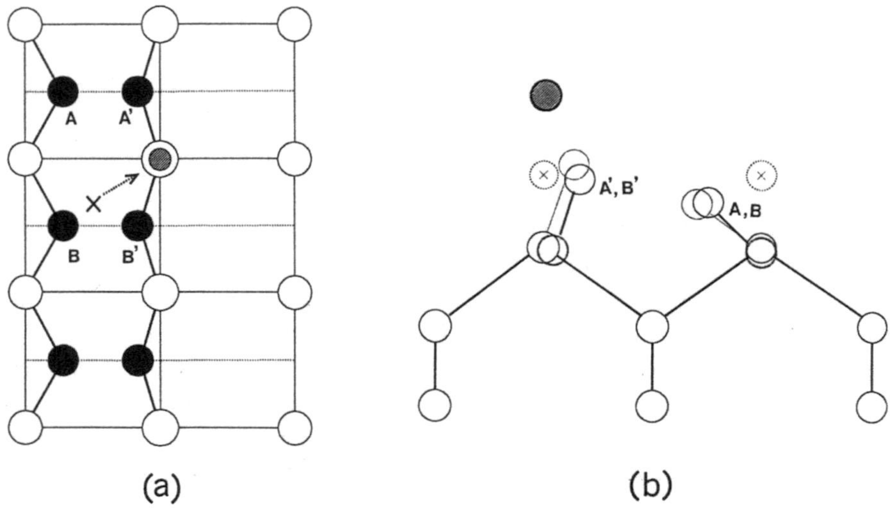

Figure 4. (a) A top view of the 2×3 surface unit cell for the 1/6 coverage of Na. The structural optimization was started by putting a Na atom at the position indicated by a cross. The Na atom moves in the direction shown by an arrow and stops at the position indicated by a hatched circle. The filled and open circles denote the substrate Si atoms in the top-most surface layer and the second layer, respectively. (b) The corresponding side view. The hatched circle denotes the Na atom. The thin dotted circles with and without a cross denote the ideal bulk terminated Si surface and the asymmetric dimer of the clean surface, respectively. Appreciable displacements can also be seen in the second layer. The Si atoms designated by A, A', B and B' correspond to those in (a).

ordering of the asymmetric Si-Si dimers around the adatom is not correctly reproduced by the present calculation. There may also be some difference between K and Na. Still it is very interesting that the present calculation does indicate a dramatic change in the preferred adsorption site between the high coverage and low coverage cases.

4. Concluding Remarks

We performed the optimization of the structure for (i) the clean Si(001)-2×1 surface and (ii) Na adsorbed Si(001) surface with $\theta = 1, 1/2$ and $1/6$ by using first-principles molecular dynamics [2,25].

We do not repeat the calculated results but would like to point out some problems in the present calculations. In the calculated optimum structures, the inter-atomic distances are appreciably underestimated. In particular, the Na-Si distance is generally underestimated by about 15%. Most of this underestimation may originate from a simplified treatment of the exchange-correlation-potential and -energy. We assumed in the present work that the core electron density and the valence electron density are spatially separated.

An improvement of this simplified treatment is rather straightforward [29,30] and the improved calcultions are under way. Some of the subtle aspects in this paper may therefore be modified in the forthcoming paper [31].

Acknowledgements

This work was partly supported by Grant-in-Aid for Scientific Research from Ministry of Education, Science and Culture. Numerical Calculations were performed at the Computer Centers of the University of Tokyo and Institute for Molecular Science.

References

1. R. O. Jones and O. Gunnarsson : Rev. Mod. Phys. **61**, 689 (1989).
2. R. Car and M. Parrinello : Phys. Rev. Lett. **55**, 2471 (1985).
3. D. Levine : Surf. Sci. **34**, 90 (1973).
4. T. Aruga, H. Tochihara and Y. Murata : Phys. Rev. Lett. **53**, 372 (1984).
5. Y. Enta, K. Kinoshita, S. Suzuki and S. Kono : Phys. Rev. B **36**, 9801 (1987).
6. T. Abukawa and S. Kono : Phys. Rev. B **37**, 9097 (1987).
7. T. Kendelewicz, P. Soukiassian, R. S. List, J. C. Woicik, P. Pianetta, I. Lindau and W. E. Spicer : Phys. Rev. B **37**, 7115 (1988).
8. H. Tochihara and Y. Murata : Surf. Sci. **215**, L323 (1989).
9. G. S. Glander and M. B. Webb : Surf. Sci. **222**, 64 (1989).
10. G. S. Glander and M. B. Webb : Surf. Sci. **224**, 60 (1989).
11. P. Soukiassian, M. H. Bakshi, Z. Hurych and T. M. Gentle : Surf. Sci. Lett. **221**, L759 (1989).

12. Y. Hasegawa, I. Kamiya, T. Hashizume, T. Sakurai, H. Tochihara, M. Kubota and Y. Murata : Phys. Rev. B **41**, 9688 (1990).
13. Y. Hashizume, I. Sumita, Y. Murata, S. Hyodo and T. Sakurai : J. Vac. Sci. Technol. A **9**, 742 (1991).
14. S. Tanaka, N. Takagi, N. Minami and M. Nishijima : Phys. Rev. B **42**, 1868 (1990).
15. C. M. Wei, H. Huang, S. Y. Tong, G. S. Glander and M. B. Webb : Phys. Rev. B **42**, 11284 (1990).
16. S. Ciraci and I. P. Batra : Phys. Rev. B **37**, 2955 (1988).
17. R. V. Kosowski and M. -H. Tsai : Phys. Rev. Lett. **60**, 546 (1988).
18. Y. Ling, A. J. Freeman and B. Delley : Phys. Rev. B **39**, 10144 (1989).
19. I. P. Batra : Phys. Rev. B **39**, 3919 (1989).
20. H. Ishida and K. Terakura : Phys. Rev. B **40**, 11519 (1989).
21. I. P. Batra : Phys. Rev. B **43**, 12322 (1991).
22. K. Kobayashi, S. Blügel, H. Ishida and K. Terakura : Surf. Sci. **242**, 349 (1991).
23. G. B. Bachelet, D. R. Hamann and M. Schlüter : Phys. Rev. B **26**, 4199 (1982).
24. L. Kleinman and D. M. Bylander : Phys. Rev. Lett. **48**, 1425 (1982).
25. A. R. Williams and J. Soler : Bull. Am. Phys. Soc. **32**, 562 (1987).
26. M. C. Payne, J. D. Joannopoulos, D. C. Allen, M. P. Teter and D. H. Vanderbilt : Phys. Rev. Lett. **56**, 2656 (1986).
27. R. M. Tromp, R. J. Hamers and J. E. Demuth : Phys. Rev. Lett. **55**, 1303 (1985).
28. R. J. Hamers, R. M. Tromp and J. E. Demuth : Phys. Rev. B **34**, 5343 (1986).
29. S. G. Louie, S. Froyen and M. L. Cohen : Phys. Rev. B **26**, 1738 (1982).
30. M. M. Dacorogna and M. L. Cohen : Phys. Rev. B **34**, 4996 (1986).
31. K. Kobayashi, Y. Morikawa, K. Terakura and S. Blügel : in preparation.

Full Potential KKR: Applicability to Simulated Annealing Model

H. Akai[1], *B. Drittler*[2], *and P.H. Dederichs*[2]

[1]Department of Physics, Nara Medical University, Kashihara, Nara 634, Japan
[2]Institut für Festkörperforschung, Forschungszentrum Jülich, W-5170 Jülich, Fed. Rep. of Germany

The full potential Korringa–Kohn–Rostoker Green's function method with implementation of various numerical techniques is shown to provide a convenient and extremely fast numerical procedure which can calculate the electronic structure of transition metals as well as non-transition metals without any specific approximation other than the angular momentum truncation in the expansion of the wave functions. The method seems especially suitable for application to the dynamical simulated annealing model.

1. Introduction

Much attention has been paid to the dynamical simulated annealing model (DSAM) invented by Car and Parrinello several years ago [1], particularly in relation to the electronic structure calculation of large systems of which the geometrical structure is not known a priori. Since this technique is most effectively used for plane wave expansions of electron wave functions, it has so far been applied mostly to semiconductors and molecules composed of typical elements. It is however strongly desired that a similar strategy can be realized also for systems containing transition metal elements, for which the plane wave expansion is totally inadequate. The main purpose of the present paper is to investigate this possibility.

The full potential KKR (FPKKR) band structure calculation [2] seems to have several favorable features as a general method to treat the electronic structure of a wide range of condensed matter systems. It is well known that for FPKKR there still remains some controversy in its basic formulation [3]. From practical points of view, however, the difficulty in FPKKR (often referred to as near-field corrections), which in essence is related to ℓ–convergence of the partial wave expansion of the cell centered Green's function, is not a big problem. Putting the convergence problem aside, we would rather take a pragmatic way to proceed as long as it does not give rise to serious drawbacks. The biggest advantage of the method lies in its computational efficiency which might be essential in combining electronic structure calculations with DSAM. Another important feature of the FPKKR is that it can calculate the force acting on nuclei accurately from the Hellmann–Feynman force, being very different from finite–basis–set approaches, which requires the so-called Pulay correction [4].

Recently Drittler et al. have developed the FPKKR–Green's function method and used it as a practical tool to calculate the electronic structure of complex systems [5-7]. They applied it to various impurity problems and calculated the force acting on the nearest neighbor atoms of an impurity. The forces obtained in this way, together with the bulk elastic data, were used to estimate the lattice relaxation around various impurities in Cu [8,9]. Their results compared well with experiments, proving the reliability of FPKKR [5-9].

Such nice features of the FPKKR method are greatly enhanced by the use of several numerical techniques. Our preliminary calculations show that it has the ability to perform a full potential band structure calculation of, e.g., Cu within local density functional theory, where each iteration cycle takes less than a second of CPU time on a usual supercomputer. All these observations indicate that FPKKR is a good candidate for combined band structure calculation and DSAM. In the following we will give a prescription to attain such a high speed for FPKKR that is necessary for combination with DSAM. The FPKKR method is briefly outlined in §2. Basic ideas due to Drittler et al., which makes FPKKR practicable, are also explained. Section 3 is devoted to how to realize the fast FPKKR procedure mentioned above. We discuss the possibility of combining it with DSAM in §4.

2. Full Potential KKR Method

In FPKKR we divide the whole crystal into the Wigner–Seitz cells which define a non-overlapping cell potential V^n in each cell. For such cellular potentials the cell centered Green's function is expanded as [3]

$$G(\boldsymbol{r}+\boldsymbol{R}_m, \boldsymbol{r}'+\boldsymbol{R}_n) = \delta_{mn} G_s^m(\boldsymbol{r},\boldsymbol{r}') + \sum_L R_L^m(\boldsymbol{r}) \sum_{L'} G_{LL'}^{mn} R_{L'}^n(\boldsymbol{r}') \quad (1)$$

where $\boldsymbol{R}^n$ denotes a center of the Wigner–Seitz cell, R_L^n and G_s^m are the single-site wave function and Green's function defined by

$$R_L^m(\boldsymbol{r}) = J_L(\boldsymbol{r}) + \int \mathrm{d}\boldsymbol{r}' g(\boldsymbol{r},\boldsymbol{r}') V^m(\boldsymbol{r}') R_L^m(\boldsymbol{r}') \quad (2)$$

and

$$G_s^m(\boldsymbol{r},\boldsymbol{r}') = g(\boldsymbol{r},\boldsymbol{r}')V^m(\boldsymbol{r}') + \int \mathrm{d}\boldsymbol{r}'' g(\boldsymbol{r},\boldsymbol{r}'')V^m(\boldsymbol{r}'')G_s^m(\boldsymbol{r}'',\boldsymbol{r}'), \quad (3)$$

where $J_L(\boldsymbol{r}) = j_\ell(\sqrt{E}r)Y_{\ell m}(\hat{\boldsymbol{r}})$ is the spherical Bessel function (with the angular part $Y_{\ell m}$) and $g(\boldsymbol{r},\boldsymbol{r}')$ is the free space Green's function. Correspondingly the cell centered expansion for the free space Green's function has a form

$$g(\boldsymbol{r}+\boldsymbol{R}_m,\boldsymbol{r}'+\boldsymbol{R}_n) = \delta_{mn}g(\boldsymbol{r},\boldsymbol{r}') + \sum_L J_L(\boldsymbol{r}) \sum_{L'} g_{LL'}^{mn} J_{L'}(\boldsymbol{r}'). \quad (4)$$

An integral equation which connects the perturbed Green's function to the unperturbed free space Green's function yields the relation between $G_{LL'}^{mn}$ and $g_{LL'}^{mn}$:

$$G_{LL'}^{mn} = g_{LL'}^{mn} + \sum_{L''m'} g_{LL''}^{mm'} \sum_{L'''} t_{L''L'''}^{m'} G_{L'''L'}^{m'n} \quad (5)$$

with the t matrix defined for each cell potential by

$$t_{LL'}^m = \int \mathrm{d}\boldsymbol{r}\, J_L(\boldsymbol{r})V^m(\boldsymbol{r})R_{L'}^m(\boldsymbol{r}) \quad (6)$$

The prescription given above is not so obvious in the case of the full potential as in the muffin-tin case. Actually Zeller showed [3] that the validity of the expression holds only when the infinite sum with respect to L is regarded as a Borel sum.

Next we outline how to solve the equations (Eqs.(1)-(6)) practically, following Drittler et al. [5-7]. First we factorize the cell potential in the form like

$$V^n(\boldsymbol{r}) = U^n(\boldsymbol{r})\Theta(\boldsymbol{r}) \quad (7)$$

where $\Theta(\boldsymbol{r})$ is a Heavyside function, being equal to 1 within the cell and zero otherwise. Since $U^n(\boldsymbol{r})$ is arbitrary outside the cell, we can choose it such that U^n extends smoothly beyond the cell boundary. These two factors are separately expanded into the spherical harmonics (or any convenient linear combinations of them), finally yielding a similar expansion for V^n:

$$V^n(\boldsymbol{r}) = \sum_L V_L^n(r)Y_L(\hat{\boldsymbol{r}}) \quad (8)$$

Such factorization is useful since the expansion coefficients (called radial shape functions) $\Theta_L(r)$ introduced by Andersen and Woolley [10] can be calculated accurately and stored beforehand by use of the numerical method developed by Stefanou, Akai and Zeller [11]. The single-site wave function R^n_L satisfying the Schrödinger equation with a single cell potential V^n also may be expanded into the spherical harmonics:

$$R^n_L(\boldsymbol{r}) = \sum_{L'} R^n_{L'L}(r) Y_{L'}(\hat{\boldsymbol{r}}) \tag{9}$$

Substituting these expansions into the single site Schrödinger equation yields coupled radial equations which $R_{L'L}$'s should satisfy for a given set of V_L 's. Usually it is rather time consuming to integrate these coupled differential equations. The method proposed by Drittler et al. instead tries to obtain the solution by solving the integral equation Eq.(2) perturbatively. Starting from ϕ_L, i.e., the solution of the radial Schrödinger equation for the spherical part $V_{L=0}$ of the potential, $R_{L'L}$ can be calculated to an arbitrary order of the Born series by use of the integral equation iteratively;

$$R^{\text{out}}_{L'L}(r) = \phi_L(r)\delta_{LL'} + \sum_{L''} \int r'^2 \mathrm{d}r' g_{L'}(r,r') \Delta V_{L'L''}(r') R^{\text{in}}_{L''L}(r') \tag{10}$$

where $g_L(r,r')$ is the L diagonal single site Green's function for $V_{L=0}$ and $\Delta V_{LL'}$ is given by

$$\Delta V_{LL'}(r') = \sum_{L''(\ell''\neq 0)} C_{LL'L''} V_{L''} \tag{11}$$

with $C_{LL'L''}$ being the Gaunt numbers in the usual notation. The superscripts "in" and "out" imply the input and output for each iteration step. Since the non-spherical part of the potential is much smaller than the spherical part, the iteration process converges rapidly. Moreover the procedure simultaneously yields the t matrix of the cell potential (Eq.(6)). These features of the method make FPKKR really practicable.

The remaining part of the FPKKR method is, except for the rather minor point that the t matrices are not diagonal anymore, very much the same as the usual KKR method for muffin-tin potentials, i.e., to solve the Dyson-like equation given by Eq.(5). The non-spherical part of the charge density, however, has to be retained to generate the non-spherical potential used in the next LSD (local density functional) iteration cycle.

3. Numerical Method

Formally the solution of Eq.(5) is expressed as

$$G = (1 - gt)^{-1}g. \tag{12}$$

Usually in KKR band structure calculations one tries to find the zeros of $\det|1 - gt|$ to obtain the dispersion relation $E(\boldsymbol{k})$. This, however, is not necessary if one only wants to know the ground state properties such as charge/spin densities, total energy, etc. A better alternative in this case is to invert $(1 - gt)$ to obtain G for some fixed energy points, which may be chosen in the upper complex half-plane. All the quantities related to the ground state are then calculated by integrating G along a complex energy contour and adding it to the contribution of the single site Green's function G_s (see Eq.(1)). This method is especially favorable when the unit cell contains a large number of atoms. The number of zeros of the determinant increases in proportion to the number of atoms within the unit cell, which clearly makes the first method more time consuming while the second one is unaffected except that the matrix size becomes bigger. The integration along the complex energy contour can be performed quite economically, e.g., 30 to 40 energy points covering an energy interval of around 1 Ry can give a highly accurate total energy.

On the other hand the contribution of the single site Green's function G_s is more easily integrated along the real energy axis. This is related to the fact that the irregular solutions (i.e., singular at the origin) of the Schrödinger equation enter only to the real part of G_s for real energies. Thus as far as the imaginary part of Green's function is concerned we can avoid calculating the irregular solutions which sometimes causes numerical problems. Note that such a separation of the Green's function, integrating them separately along different paths, is only possible since both the total and the single site Green's function in our definition are true Green's functions, and hence, analytic for complex energies.

To solve the Schrödinger equation for complex energies is not favorable from numerical points of view. Instead, we can solve them for some real energies (typically 7 to 10), construct appropriate polynomials to fit them and make an analytical continuation to the complex plane. Actually we need not the polynomials which fit the radial wave functions on each radial mesh point, but ones which fit the information about the t matrices. This procedure finally gives the weights to the radial wave functions calculated for a set of fixed real energies. The charge/spin densities are then expressed as weighted sums of the contributions of the wave functions for these energies. We refer to reference [12] for further details.

The structural Green's functions $g_{LL'}(\boldsymbol{k})$ (the Fourier component of $g^{mn}_{LL'}$) are rather easily treated. However, the computational efficiency is greatly enhanced by calculating them for a set of energies and k points beforehand and fitting (the non-singular part of) them with energy polynomials. After such manipulation, the most time consuming part might be the matrix inversion given in Eq.(12). Fortunately this part can be fully vectorized, not in the sense of a usual vectorization, i.e., with respect to columns or rows of the matrix, but with respect to the energies and the k points; each $(E, \boldsymbol{k})$ contributes independently and furthermore completely the same treatment for all these points is allowed. Thus the assembly of $(E, \boldsymbol{k})$ forms a long vector the length of which is the number of energy points (30 to 40) time the number of k points (depends on the number of atoms in the unit cell). In our experience, as small as 10 k points already is enough to attain an acceleration of a factor 50 to 100 (compared with the same procedure before the vectorization) on typical supercomputers (CRAY, NEC and HITAC are tried).

4. Results and Discussion

We tested the FPKKR procedure with implementation of the numerical techniques mentioned in the previous section; fcc Cu is chosen as a typical test case and the efficiency and reliability of the method are examined. Table I compares the results of various band structure calculations all performed for fcc Cu. In the table, "KKR1" denotes the usual KKR band structure calculation which searches the zeros of the KKR matrix while "KKR2" exploits the method of the direct integration of Green's

Table I. Comparison of various KKR schemes applied for fcc Cu. KKR1, KKR2 and FPKKR denote the standard KKR, KKR with direct integration of Green's function (plus some numerical technique mentioned in the text), and the full potential KKR, respectively; E_{total} denotes the calculated total energy (in Ry), around 40 iterations have been made to attain LSD convergence in these calculations; CPU indicated the CPU time (in sec) needed for each LSD iteration cycle; KKR is performed on NEC-ACOS2000, others on HITAC-S820.

	KKR1	KKR2	KKR2	FPKKR
ℓ_{max}	2	2	3	3
E_{total} (Ry)	-3275.7610	-3275.7606	-3275.7661	
CPU (sec)	18	0.12	0.43	0.56

function used in FPKKR but applied to the muffin-tin potential case. We used a normal computer for the KKR calculation since it has not been coded such as to run efficiently on a supercomputer. However we feel that it is rather difficult to get a significant improvement by vectorizing the usual KKR band structure code. For all these calculations, 240 k points in the irreducible wedge of the first BZ are used. For comparison, we tried out two cases for the fast KKR (KKR2), one with $\ell_{\max} = 2$ and the other with $\ell_{\max} = 3$. For the FPKKR $\ell_{\max} = 3$, at least, is necessary since otherwise the wave function in the interstitial region is too poorly described; this however is not the case for the muffin-tin KKR where the wave functions are calculated only inside the muffin-tin sphere and the charge accumulated in the interstitial region is obtained from the difference between the total number of states below the Fermi level and the charge inside the sphere.

The remarkable features seen in the table are: i) the substantial improvement of the fast KKR (KKR2) over standard KKR (KKR1) and ii) the fact that FPKKR does not need much more time compared with KKR. It however might not be quite fair to compare KKR1 running on a normal computer with KKR2 on a supercomputer. We only want to stress that the KKR procedure can be extremely fast if a sufficient numerical elaboration is made. This may be one of the biggest advantages of KKR over other schemes. The increase of the CPU time when we go from $\ell_{\max} = 2$ to $\ell_{\max} = 3$ looks reasonable. Though we are rather reluctant to go beyond $\ell_{\max} = 3$ at present, it would surely be necessary in the cases of open crystal structures or structures with much lower symmetry.

In the following we will examine the possibility of applying FPKKR to DSAM. Contrary to the usual Car–Parrinello procedure, our method has to deal with the electronic structure part and the molecular dynamics part separately, because no wave function appears explicitly in FPKKR, which otherwise would provide new degrees of freedom for the molecular dynamics part. This is both an advantage and a disadvantage. The size of the matrix we have to deal with is in this way greatly reduced; 16×16 may be compared with the plane-wave-basis cases which may require more than ten times as many basis functions even though some appropriate augmentation to the wave functions are made. On the other hand we miss some attractive techniques related to the electronic structure calculation, which are typical for the Car–Parrinello procedure [1].

First we check whether the computational speed attained by the FP-KKR is realistic for DSAM or not. The main difference between the present calculation and that needed for DSAM lies in the number of atoms N in the unit cell. The steps required for the numerical procedure increases, roughly speaking, as N^2 in the present scheme. Another factor we have to take into account is the lack of the symmetry in the case of DSAM. This increases the number of the non-zero radial wave functions $R_{LL'}$'s from 8 for the present case to 136 for a general case ($\ell_{\max} = 3$). Fortunately the procedures which handle these wave functions are not very time consuming and moreover increase only linearly with N, and hence, are negligible for large N (say $N \sim 10$). Also one has to take account of the radial shape function $\Theta_L(r)$ (see Eq.(7)), which may be tabulated for the certain symmetry cases (fcc, bcc, ...) but not for general cases. What we have to do is to construct the Voronoi cells and calculate the shape functions for each polyhedron. An efficient way to calculate the shape functions for such general cases has recently been developed by Stefanou and Zeller [13] which, combined with algorithms for the automatic construction of the Voronoi cells [14-16] for a given atomic arrangement, will provide us with a convenient setting up for the procedure. This also contributes to the CPU time linearly with respect to N. Our estimation of the CPU time needed for a single run for FPKKR is then around a hundred seconds for ~ 10 atoms by use of the now available supercomputers; this figure seems substantial, but not prohibitive for the molecular dynamics.

Next we look into the way to realize molecular dynamics. Since the dynamics with respect to the wave functions does not enter, the only thing that has to be considered is the force acting on the nuclei. An important question is whether we will have to consider the Pulay force or not in the present scheme. This is related to how completely the wave functions are given, either implicitly or explicitly. In FPKKR we use the energy dependent wave functions which exactly satisfy the Schrödinger equation within each cell (Wigner–Seitz or Voronoi) and include infinite number of plane waves to describe the multiple scattering caused by the cell potentials. In this case the extra force, which might contribute in addition to the Hellmann–Feynman force, can occur only from incompleteness in the L expansion. The problem, therefore, is not serious: if we confirm that $\ell = \ell_{\max}$ is enough to describe the electronic structure of the system, using $\ell = \ell_{\max} + 1$ would be enough to make the Pulay force vanishingly small (more precisely, negligible in the same approximation as for the total energy).

In conclusion, the combination of the FPKKR and the DSAM is sufficiently realistic and looks promising in the sense that it might enable us to deal with metallic systems composed of transition metal elements in a way analogous to the one which clearly has been a breakthrough in semiconductor physics.

Acknowledgement

The authors would like to express sincere thanks to Drs. N. Stefanou and R. Zeller for valuable discussions.

References

[1] R. Car and M. Parrinello, Phys. Rev. Lett. **55**(1985)2471.

[2] A.R. Williams and J. van W. Morgan, J. Phys. C: Solid State Phys. **5**(1972)l293; ibid. **7**(1974)37.

[3] R. Zeller, J. Phys. C: Solid State Phys. **20**(1987)2347.

[4] See, e.g., P. Pulay, Mol. Phys. **21**(1971)329.

[5] B. Drittler, Ph. D. Thesis, Tech. University Aachen (1990).

[6] B. Drittler, M. Weinert, R. Zeller and P.H. Dederichs, Phys. Rev. **42**(1990)9336.

[7] B. Drittler, M. Weinert, R. Zeller and P.H. Dederichs, Solid State Commun. (1991), in press.

[8] K. Abraham, Diploma Thesis, Tech. University Aachen (1990).

[9] P.H. Dederichs, T. Hoshino, B. Drittler, K. Abraham and R. Zeller, Phisyca B (1991), in press.

[10] O.K. Andersen and R.G. Woolley, Mol. Phys. **26**(1973)905.

[11] N. Stefanou, H. Akai, and R. Zeller, Comp. Phys. Commun. **60**(1990)231.

[12] H. Akai, J. Phys.: Condens. Matter **1**(1989)8045.

[13] N. Stefanou, R. Zeller, J. Phys: Condens. Matter (1991), in press.

[14] M. Tanemura, T. Ogawa and N. Ogita, J. Comput. Phys. **51**(1983)191.

[15] N.N Medvedev, J. Comput. Phys. **67**(1986)223.

[16] R. Riedinger, M. Habar, P. Oelhafen and H.J Günterodt, J. Comput. Phys. **74**(1988)61.

Application of the Car-Parrinello Molecular Dynamics to Some Microclusters

C. Satoko

Department of Physics, College of Humanities and Sciences,
Nihon University, Sakurajousui, Setagaya-ku, Tokyo 156, Japan

1. Introduction

In a calculation of molecular dynamics we have used the various interactions between the atoms. The simplest interaction is written as the summation of the pair interaction, which is monotonically dependent on the pair distance, i.e., has only one minimum, at the equilibrium position. In such an interaction the stable structure has a close packed structure to lower the total energy. In some systems such as covalent materials a three-body interaction is necessary for the description of the stable structure, which is not the close packed structure. It consists of interactions with the fixed parameters conformed to experiments, such as the Morse potential, Lennard-Jones potential, Stillinger and Weber potential and so on [1]. Currently, these parameters are determined by conforming them to the ab-initio calculated potentials. Almost all molecular dynamics of solid and liquid systems have been investigated by this method.

The pair interaction is good for solid and liquid systems. These systems are isotropic. However, these interactions are questionable for the total energy calculation of microclusters, because microclusters are not isotropic around the surface. The bond properties around the surface region are different from the interior region. Therefore, these simple interactions are not adequate to determine the correct geometries around the surface region and interior bulk region. So, the second category is one calculated by an interaction with variable parameters dependent on the surroundings. For example, in the case of the Chelikowsky potential, one parameter is the number of dangling bonds in close vicinity to a given atom [2]. However, the interaction form obtained is very complex.

The third category is one calculated by the electronic structures. This method is general, but requires much time to calculate the total energy. Nowadays, time is saved by using high speed computers and efficient techniques such as the Car-Parrinello method [3]. Most of the work is done on the basis of plane waves due to the use of the fast Fourier transform method and no correction term of the Hellmann-Feynman force.

In this paper, I would like to discuss dynamical calculation using the basis of the atomic orbitals. This is suitable for the analysis of the chemical processes such as reaction, bonding, dissociation and evaporation. In addition, it is also used for the localized orbital such as the d-orbital. However, it uses much computer time to calculate the correction term of the Hellmann-Feynman force

[4]. We are trying to overcome these problems and to calculate the geometries and dynamics of microclusters.

There are many methods for electronic structure calculation in the LCAO approximation. A simple method consists of using parameter model methods such as the tight binding model. The second one is the modification of the LCAO-Local density functional-force methods in order to save computer time. The third one is the charge density calculation such as the Thomas-Fermi method. In the next section, I will discuss the application of the tight binding models to potassium clusters, silicon clusters and carbon clusters.

2. Calculation method

2.1 Tight-binding Hamiltonian Model

We calculate the equilibrium structures of some microclusters by using the following Hamiltonian:

$$H = H_e + H_c + H_r \quad . \tag{1}$$

The first term is the transfer energy of the electrons between the atoms:

$$H_e = \sum_{ij\sigma} t_{ij} c^+_{i\sigma} c_{j\sigma} \quad , \tag{2}$$

where c^+ and c are, respectively, the creation and annihilation operators of the valence electrons. We study the microclusters of the potassium atom with the s valence electron and the silicon and carbon atoms with the s and p valence electrons. The second term is the intra-atomic Coulomb interaction between the electrons

$$H_c = \sum_{i\sigma} U_i c^+_{i\sigma} c_{i\sigma} c^+_{i\bar{\sigma}} c_{i\bar{\sigma}} \quad . \tag{3}$$

The final term is the repulsive interaction written as the summation of the pair potential:

$$H_r = \sum_{km} \phi_{km} \quad . \tag{4}$$

We assume the following two functions of the internuclear distance r for the bond integrals and repulsive potentials : the first one is the exponential damping dependence [5,6]:

$$t_{ij} = t_0 e^{-\frac{r_{ij}}{\tau}} \quad . \tag{5}$$

The repulsive potential is also an exponential damping potential.
The second one is the reciprocal power damping dependence [7,8]:

$$t_{ij} = t_0 \frac{1}{r_{ij}^{\,l_i + l_j + 1}} \tag{6}$$

This power is the sum of the orbital angular momentums of the related atoms plus one , for example, the bond integrals for s-s interactions vary as 1/r. The repulsive potential is also a reciprocal power damping dependence. The repulsive power is usually taken as two times the power of the tight binding parameter. Then the power value of the alkali-atoms with s-orbitals is two. However, since the range of the repulsive potential is too long, the stable structure of the three potassium atoms becomes linear, which does not fit to the observed isosceles triangle. Therefore, the power is determined to get the correct geometry of potassium clusters K_3 and we change the power to 12.

2.2 Dynamical method for the search for the energy minimum points

We find the minimum point over the adiabatic potential surface by using the Car-Parrinello dynamical equation method. Total energy E is given by the atom positions R and the wave functions which are given by the linear combination of the atomic orbitals:

$$\psi_i(\mathbf{R}) = \sum_{\alpha} C_{i\alpha}(\mathbf{R})\, \chi_\alpha(\mathbf{R}) \tag{7}$$

The suffix i is used for the eigenfunction of the Hamiltonian equation (1) and the suffix α is used for the atomic orbitals. The vector R represents the atom coordinates which are given as functions of time t. When the total energy is minimized with respect to the coefficients of equation (7), these functions are the eigenstates.

The dynamical method uses the following dynamical equations to minimize the total energy with respect to both the atom positions and the wavefunctions:

$$m\ddot{\mathbf{R}}_\alpha = -\frac{dE(\mathbf{R},\psi)}{d\mathbf{R}_\alpha}, \tag{8}$$

$$\mu \ddot{C}_{i\alpha} = -\sum_{\beta} h_{\alpha\beta} C_{i\beta} + \sum_{j} \varepsilon_{ij} C_{j\alpha}, \tag{9}$$

where ε_{ij}'s are the Lagrangian undefined multipliers used to conserve the norm of the wave functions. If the value of this equation (9) is zero, then the wave function is a eigenfunction.

The friction terms are added to these dynamical equations to slow down the velocity of the atoms and to gradually stop the atoms at the minimum point. As the minimum point is not always the lowest minimum point of the total energy , we must repeat this calculation with different initial conditions. Finally, we shall obtain the true eigenfunction and the stable geometries in the ground state.

3. Static results

We will study the stable structure of potassium, silicon and carbon clusters and determine the magic numbers in related to the cluster formation.

3.1 Potassium clusters

As is well known, the magic numbers of alkalimetals 2, 8, 18, 20 and so on, which are observed by experiments, are determined theoretically by using the jellium model [9,10]. The jellium model assumes the continuum and round shape of the microcluster so it is difficult to study the reaction mechanism.

We have tried to study the calculated total energy as a function of the number of atoms in the case of the exponential damping parameters. However, the K_8 cluster is no more stable than the others. I cannot get the magic number of the potassium clusters [11]. Next, I calculate the stable structure by using the reciprocal damping parameters.

Figure 1 shows the calculated stable structures of potassium clusters. The strucutres in figure 1 correspond to 3, 4, 5, and 8 atoms. These geometries are similar to the ab-initio calculated results. The geometry of the eight atoms is a distorted cube. The geometry of the thirteen atoms is an icosahedron.

To compare the order of the cluster fragmentation, I plot the fragmentation energy which is defined as the energy difference between E_n and E_{n-1}. The fragmentation energies in figure 2 show the maxima at atom numbers eight, seventeen, eighteen and twenty which are more stable than the other numbers. The magic numbers , eight, eighteen and twenty, agree with the experiments but 17 does not agree . We are now checking whether the total energy of the large cluster may be at the local minimum.

Consequently, these results show that it is possible to understand the magic numbers in the LCAO method as well as in the jellium model.

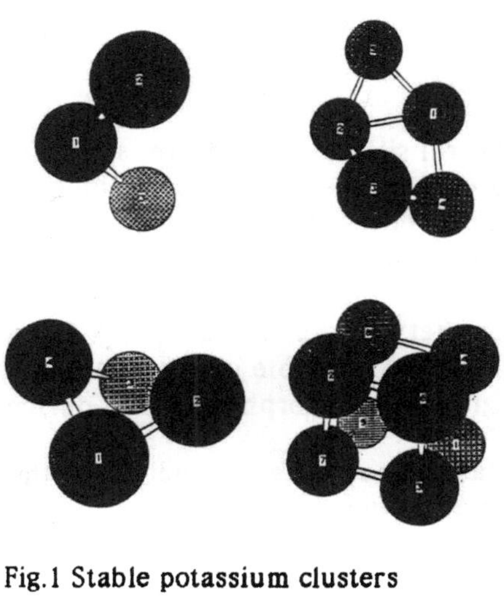

Fig.1 Stable potassium clusters

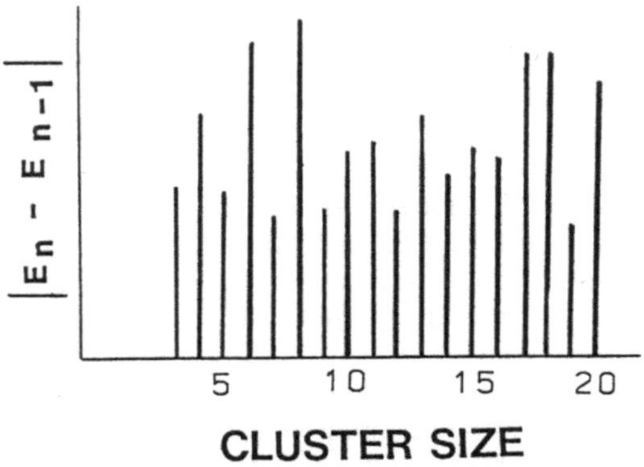

Fig.2 Fragmentation energy of K clusters

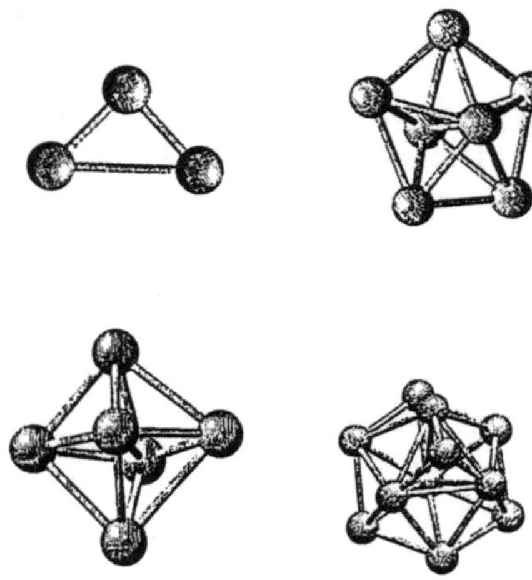

Fig.3 Stable Si clusters

3.2 Silicon clusters

The optimized stable geometries of several silicon clusters are shown in figure 3. These are very different structures from the solid. The minimum distance of each cluster increases with the size of the cluster. The structure of the Si_6 distorted octahedron cluster is different from that of the sp^3 bond. Triangles contract by 20% in length as compared to those in the crystal. This contraction is attributed to the interaction between surface dangling bonds which are now written in this figure [12,13]. The stable cluster Si_7 has a distorted-pentagon structure which may grow to an amorphous silicon during crystal growth.

Figure 4 shows the calculated fragmentation energy as a function of the number of silicon atoms. The large fragmentation energies occur in the clusters, Si_4, Si_6, Si_{10}. The magic numbers from experiments are 4, 6, and 10 [14-17]. The calculated results agree with the observed stable Si clusters.

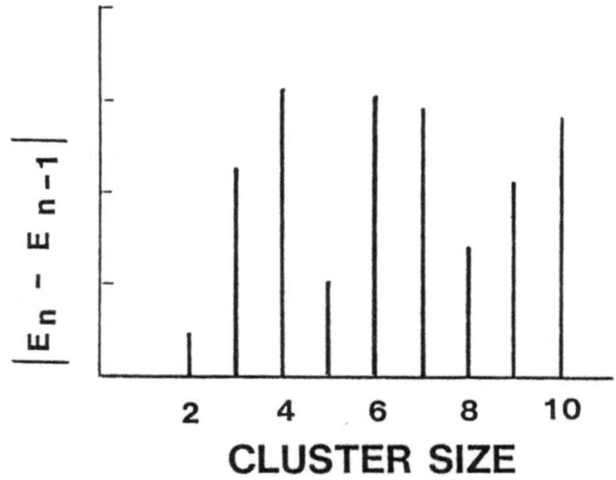

Fig.4 Fragmentation energy of Si clusters

3.3 Carbon clusters

Figure 5 shows the stable geometries of several carbon clusters. The C_3,C_4,C_5, and C_6 clusters are linear. The dependence of the bond length on the size of the clusters shows the Peierls distortions. The planar structure is more stable than the linear structure for carbon clusters larger than 7. The carbon atoms in the ground state grow in a rugged planar shape and tend to form a shape like a graphite structure. Figure 6 shows the fragmentation energy dependent on the the number of carbon atoms. The large fragmentation energies occur for 3,6, and 10 atoms. However, these numbers do not agree with observed results of laser evaporation [18,19]. This shows that the fragmentation mechanism is not different from the dissociation of a single carbon atom.

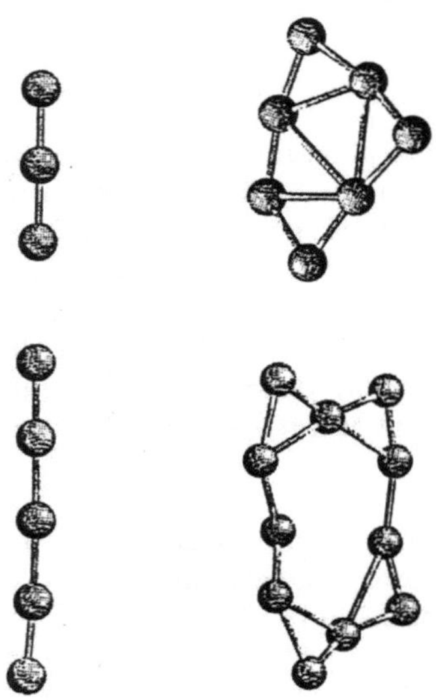

Fig.5 Stable carbon clusters

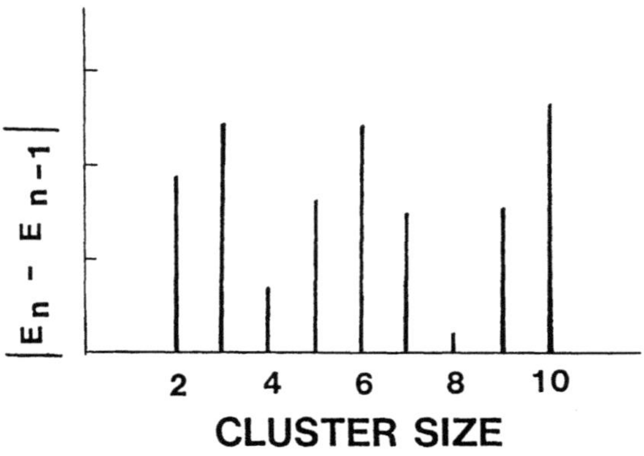

Fig.6 Fragmentation energy of C clusters

4. Dynamical results and their problems

The Car-Parrinello equations are used for calculating the dynamics of microclusters [20]. However, the dynamical trajectory does not always travel along the adiabatic surface. To check whether the dynamical trajectory moves along the exact adiabatic potential surface, we compare the results of the Car-Parrinello dynamical equation method with the exact diagonalization method. In each time step of the molecular dynamics, we need to look for the eigenfunctions of the Hartree-Fock equation (1) in the LCAO approximation. Exact diagonalization is done by the Jacobi method which transforms the matrix elements until all of the off-diagonal elements become zero. When the difference between the time step is very small (about 10-15 sec), the change of the coordinates of the atom positions is also small. The Hamiltonian matrices are approximately solved by the perturbation methods except for the great change in the wave functions. With this method, every trajectory moves along the true adiabatic potential surface.

4.1 Calculated results

For example, I will show the calculated results of potassium clusters K_3 in figure 7. The top one is a stable structure of K_3. When the total energy becomes a little bit higher, the cluster begins to vibrate in the bending mode. The figures on the left are multiple exposures of many time steps calculated by the Car-Parrinello dynamical equation. On the other hand, the figures on the right are calculated by the exact diagonalization method. When the kinetic energy is lower than the threshold energy, the left figures are the same as those on the right. However, when the kinetic energy progressively increases, the difference is apparent between the two.

Figure 8 shows the temperature T versus total energy E per atom of K_3 as determined by using both types of calculation. At lower temperatures, the T-E curve of the Car-Parrinello method is nearly identical to the T-E curve of the exact diagonalization method. This means that the dynamical trajectory is moving along the adiabatic potential curve. Above this point, the temperatures

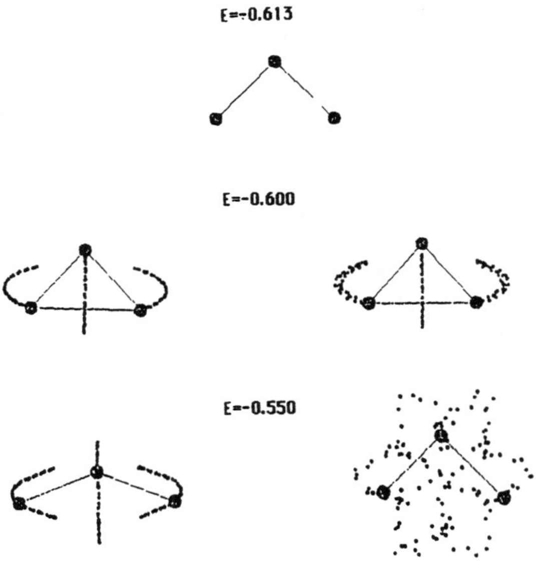

Fig.7 Multiple exposures of K_3

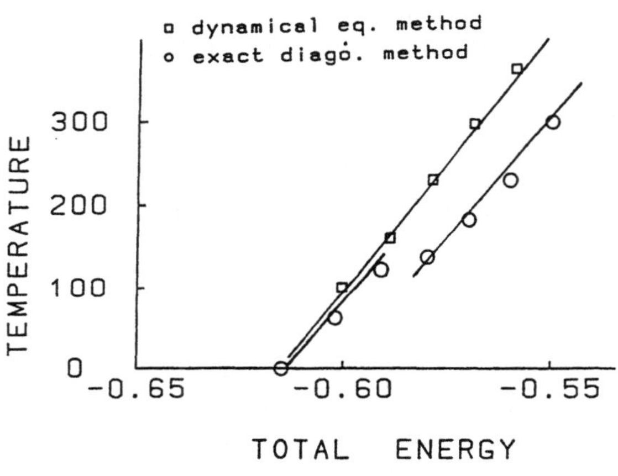

Fig.8 temperature vs energy per atom

of each do not coincide. The T-E curve of the diagonalization shifts over. The T-E curves of the K_8 cluster show the difference at increasing total energies between the methods. Next, let us consider the reason for the difference between them.

4.2 Discussion of the potential surface

The Car-Parrinello method is also used for calculating the dynamics at finite temperatures or in an excited state. Since the method solves the potential surface only along the actual dynamical trajectory, this can be applied to the dynamics of a large system such as surfaces and large microclusters.

When there is only one potential surface of the system, the dynamical trajectory is always moving along the true adiabatic potential surface. However, some potential surfaces usually cross with other potential surfaces. The symmetry of a system is not low, especially, in the case of microclusters. The potential surface at some atom positions is crossing with other potential surfaces, because their potential surfaces have different representations or degenerate representation. The trajectory passes the crossing point at some point in time and afterwards it is not easy to determine which potential surface the trajectory is moving along. To avoid this problem, the dynamical equation method follows the diabatic surface, while the diagonalization method goes along the adiabatic surface. This is the reason why the trajectories between the dynamical equation method and the exact diagonalization method are different.

When there are multi-potential surfaces, we must clarify which potential surface the dynamical trajectory is moving along. Whether the potential surface is crossing or not depends on how one solves the Hamiltonian. For example, when one solves the local density functional Hamiltonian in the restricted Hartree-Fock approximation, the obtained potential surface may be crossing with the other potential surface. The wave functions in the restricted approximation are assumed to have a higher symmetry. Therefore, if one solves the Hamiltonian in the exact method, these potential surfaces may not cross each other. In this case the trajectory should move along the adiabatic potential surface.

This problem may be related to the search for the minimum energy over the potential surface in the ground state. There are many local minima over the potential surface and many crossing points between the multi-potential surfaces. This is apt to occur in the calculation of a gapless or small gap system. However, this crossing problem can be solved as follows: The Car-Parrinello dynamical method is used to calculate the wave function and geometry in the ground state. As the wave function drastically changes at discontinuous crossing points, the convergence speed of the occupation number decreases and sometimes the occupation number begins to oscillate. To avoid this decelerated convergence, the occupation at the initial geometry is assumed to be in the Fermi-distribution of some virtual finite temperature. The virtual temperature for the dynamical processes is lowered gradually to zero near the lowest minimum point of the potential surface.

5. Improved LCAO methods

The calculation of the tight binding Hamiltonian model is useful to save computer time. However, it is not easy to choose reasonable parameters. I am also trying the calculation by using the LCAO-local density functional-force method. The force acting on an atom is written as follows [4]:

$$F_\mu = -\frac{dE}{dR_\mu} = F_\mu^{HF} - \sum_i^{occ} \left\{ \int \frac{\partial \phi_i}{\partial R_\mu} (h - \epsilon_i) \phi_i \, dv + c.c \right\}. \qquad (10)$$

The first term is the Hellmann-Feynman force and the second is the correction term. The calculation of the correction term requires much computer time for a large system. The order of the correction term is very dependent on the basis set in the LCAO approach. If we use the additional basis set, the correction term becomes small. For example, the simple additional basis set is chosen as the radial derivative of the original basis set. This improvement can reduce the integration time of the correction term.

Finally we propose the optimized charge density calculation method as well as the Gordon-Kim method. As the atomic charge density is written as a Slater-type function, most parts of the integral are done by analytical forms. This approximation is good for many materials.

6.Conclusion

In summary, we have discussed the application of the dynamical equation method in the LCAO approximation. We apply this method to the structure analysis of potassium, Si and C clusters by using the Car and Parrinello dynamical equation method. To determine the magic number of the alkali metal clusters, we show that it is important for the hopping integral to vary from the inverse internuclear distance. In addition, the dynamical trajectory by the dynamical equation method does not always go on the adiabatic potential surface but on the diabatic potential surface. Therefore, in the future, I would like to discuss how the trajectory moves along the potential surface. Finally I discussed the improved LCAO method to save computer time. I hope to expand the possibility of the LCAO approach in dynamical calculations.

References

1. F.H.Stillinger and T.A.Weber:Phys.Rev.B31,5262(1985).
2. J.R.Chelikowsky,J.C.Phillips,M.Kamal,and M.Strauss: Phys.Rev.Lett.62,292(1989).
3. R.Car and M.Parrinello:Phys.Rev.Letters 55,2471(1985).
4. C.Satoko:Phys.Rev.B30,1754(1984).
5. R.P.Gupta:Phys.Rev.B 23,6265(1981).
6. F.Ducastelle and F.Cyrot-Lackmann:J.Phys.Chem.Sol.32,285(1971).
7. W.A.Harrison: Phys.Rev. B 41,6008(1990).
8. O.K.Andersen,W.Close,and H.Nohl:Phys.Rev.B17,1209(1978).
9 W.D.Knight,K.Clemenger,W.A.Heer,W.A.Saunders,M.Y.Chouand and M.L.Cohen:Phys.Rev.Letters 52,2141(1984).
10. I.Katakuse:in Microclusters ed. by S.Sugano,Y.Nishina and S.Ohnishi p10(Springer-Verlag 1987).
11.C.Satoko:ibid. p142.
12. S.Saito,S.Ohnishi,C.Satoko and S.Sugano:J.Phys.Soc. of Japan 55,1791(1986).
13. S.Saito,S.Ohnishi,C.Satoko and S.Sugano:Z.Phys.D14,237(1989).
14. L.A.Bloomfild,R.R.Freeman,W.L.Brown:Phys.Rev.Lett.54,2246(1985).
15. T.P.Martin,H.Schaber:J.Chem.Phys.83,855(1985).
16. Q.L.Zhang,Y.Liu,R.F.Curl,F.K.Tittel andR.E.Smalley: J.Chem.Phys.88,1670(1988).

17 J.R.Heath, Yuan Liu, S.C.O'Brien, Qing-Ling Zhang, R.F.Curl, F.K.Tittel and R.E.Smalley:J.Chem.Phys.83,5520(1985).
18. P.P.Radi,T.L.Bunn,P.R.Kemper,M.E.MolChan and M.T.Bowers: J.Chem.Phys.88,2809(1988).
19. E.A.Rohlfing,D.M.Cox and A.Kaldor:J.Chem.Phys.81,3322(1984).
20. P.Ballone,W.Andreoni,R.Car and M.Parrinello: Phys.Rev.Letters 60,271(1988).

Part VI

Nonequilibrium Molecular Dynamics

Response Theory, Lyapunov Instability and Rheology

D.J. Evans[1], *E.G.D. Cohen*[2], *and G.P. Morriss*[3]

[1]Research School of Chemistry, Australian National University, Canberra, ACT 2601, Australia
[2]Rockefeller University, 1230 York Ave., New York, NY 10021, USA
[3]Theoretical Physics Department, University of NSW, Kensington, NSW 2033, Australia

Abstract. We give a very brief outline of the theory and algorithms used in performing nonequilibrium molecular dynamics simulations of the properties of liquids undergoing planar Couette flow. We then give a brief account of some of the results that can be obtained using this type of calculation.

1. Algorithms and Theory

The system that we wish to consider is a thermostatted N-body system subject to a steady external field that is turned on at time t=0. Consider the ensemble-averaged transient response of an arbitrary phase variable, $B=B(\mathbf{q}_1,\mathbf{q}_2,\ldots,\mathbf{q}_N,\mathbf{p}_1,\mathbf{p}_2,\ldots,\mathbf{p}_N)$.

Mathematically, the interaction of the external field, F_e, with the system is given by the equations of motion (see figure 1). We do not assume that these equations of motion are derivable from a Hamiltonian. We add a thermostatting term, $-\alpha\mathbf{p}_i$ to the equations of motion for the momentum. This thermostat removes the heat produced in the system by the external field. We assume that we can write the equations in terms of 'peculiar momenta', $\mathbf{p}_i$. Peculiar momenta and velocities are measured relative to the local streaming velocity of the fluid.

The mathematical form of the thermostat can be derived from Gauss' Principle of Least Constraint [1]. If one fixes the total peculiar kinetic energy, $(\alpha=\boldsymbol{\Sigma}(\mathbf{F}_i+\mathbf{F}_e\bullet\mathbf{D}_i)\bullet\mathbf{p}_i/\boldsymbol{\Sigma}\mathbf{p}_i^2)$, the system is termed isokinetic and the kinetic temperature, $T_k=\boldsymbol{\Sigma}\mathbf{p}_i^2/3Nmk_B$, of the system will be constant. One can also choose α such that the internal energy of the system, $H_0=\boldsymbol{\Sigma}\mathbf{p}_i^2+\boldsymbol{\Phi}$, is a constant of the motion.

In 1985 two of us (GPM and DJE) [2] developed an exact analytical expression for the nonlinear thermostatted response in terms of a time integral of so-called Transient Time Correlation Functions (TTCF). We showed that

$$<B(t)>=<B(0)>-\beta F_e\int_0^t ds<B(s)J(0)> \tag{1}$$

where the dissipative flux, J, is defined in terms of the work done, $dH_0/dt)^{ad}$, on the system by the applied external field, F_e. The time dependence on both sides of this equation is generated from the thermostatted, field dependent equations

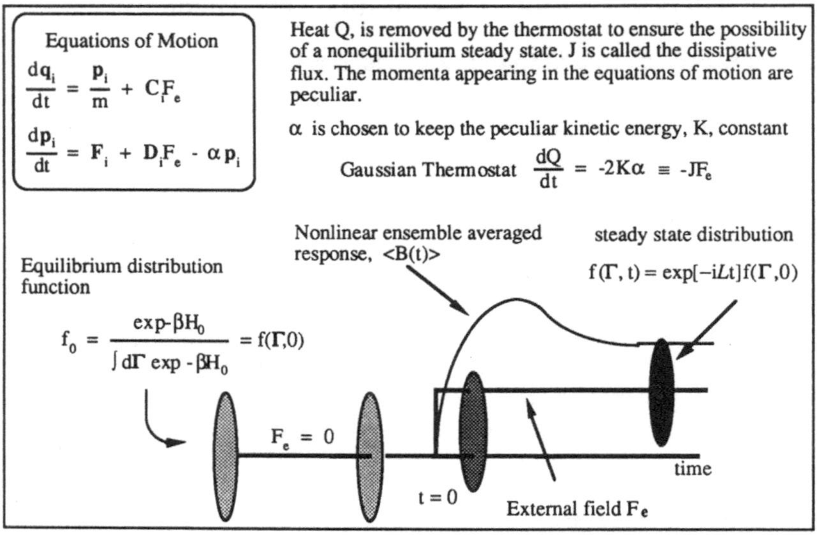

Figure 1. The thought experiment for response theory.

of motion. This formula is exact arbitrarily far from equilibrium and it can be used for thermostatted, isokinetic, isoenergetic or adiabatic (α=0), systems.

In the same paper [2] Morriss and Evans showed that the nonlinear thermostatted response could also be written in a form similar to one due to Kawasaki and Yamada [3],

$$< B(t) >=< B(0) e^{-\beta F_e \int_0^t ds J(-s)} > / < e^{-\beta F_e \int_0^t ds J(-s)} >. \qquad (2)$$

This is known as the renormalised Kawasaki form. The 'bare' Kawasaki form which was originally proposed by Kawasaki and Yamada, is identical to that given above except that the denominator is unity. It is straightforward to show that the TTCF and renormalised Kawasaki representations of <B(t)> are formally equivalent [1,2].

In the linear regime close to equilibrium, the TTCF and the bare and renormalised Kawasaki expressions for the nonlinear response all reduce to the famous Green-Kubo expressions [4]. They take the form,

$$\lim(F_e \to 0) \quad < B(t) >=< B(0) > -\beta F_e \int_0^t ds < B(s) J(0) >_0 . \qquad (3)$$

In this equation $< \ldots >_0$ denotes an equilibrium time correlation function (ie F_e=0). Since one can prove that in the thermodynamic limit, thermostatting does not change equilibrium time correlation functions [1], we do not need to specify whether thermostatting is used in the linear response regime.

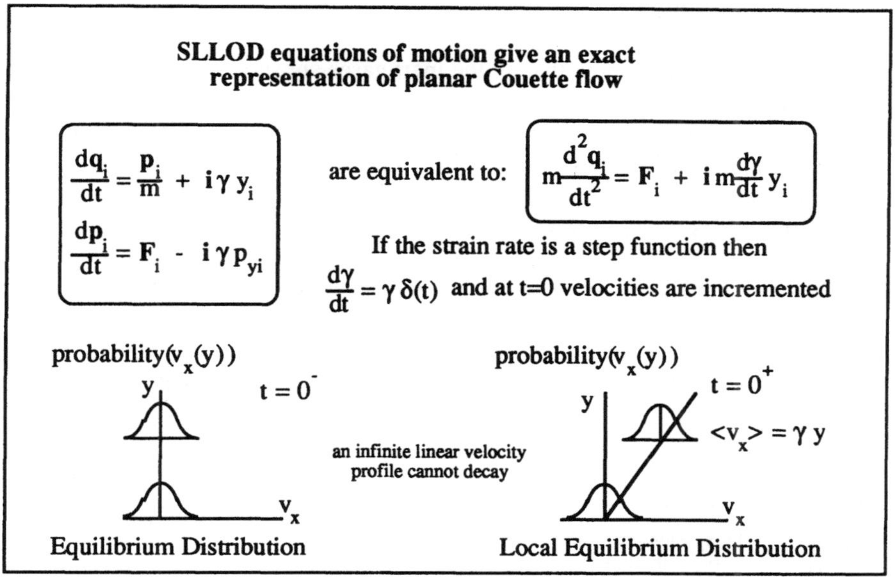

Figure 2. SLLOD equations of motion for Couette flow.

In order to apply mechanical response theory to thermal transport processes such as shear flow we need a way of incorporating the boundary conditions which drive shear flow into the equations of motion as though they are a mechanical perturbation. This is accomplished using the SLLOD equations [1] depicted in figure 2. In these equations of motion the momenta $\mathbf{p}_i$ are measured with respect to the constant linear velocity gradient which is initially imposed. If the Reynolds number is sufficiently small, the velocity profile will remain linear. It is easy to show from the SLLOD equations of motion that if $F_e=\partial u_x/\partial y=\gamma$, the strain rate, then $dH_0/dt)^{ad}=-P_{xy}\gamma V$, where P_{xy} is the xy-element of the microscopic pressure tensor, and V is the volume of the system.

In order to use the SLLOD equations of motion in computer simulations, they must be supplemented by Lees-Edwards periodic boundary conditions [5]. Under Lees-Edwards periodic boundary conditions a shearing flow is induced in the primitive cell by the motion of the periodic images above and below the primitive cell. Thus, while the lab velocities $d\mathbf{q}_i/dt$, of the particles and their images are different from one another, the SLLOD momenta, $\mathbf{p}_i$, of the particles, being peculiar momenta are identical for all their periodic images. Lees-Edwards periodic boundary conditions ensure that the system is spatially homogeneous.

2. Numerical Results

Numerical simulations show that all the inert gases are shear thinning, positively shear dilatant, viscoelastic fluids which exhibit normal stress (Weissenberg), effects. Figures 3 and 4 [1] show that the agreement between direct simulation and the computer calculated TTCF expression for the shear stress and the shear induced change in the hydrostatic pressure of a fluid composed of 72 soft discs ($\phi_{ij}=4\varepsilon(\sigma/r)^{12}$), is excellent. We use reduced units throughout this paper (ie $m=\sigma=\varepsilon=1$).

Because the Kawasaki expressions require taking averages of exponentials of integrals that grow essentially linearly with N and t, the Kawasaki formalism is not useful from a direct computational point of view. However, from a theoretical perspective, the Kawasaki expression is very important.

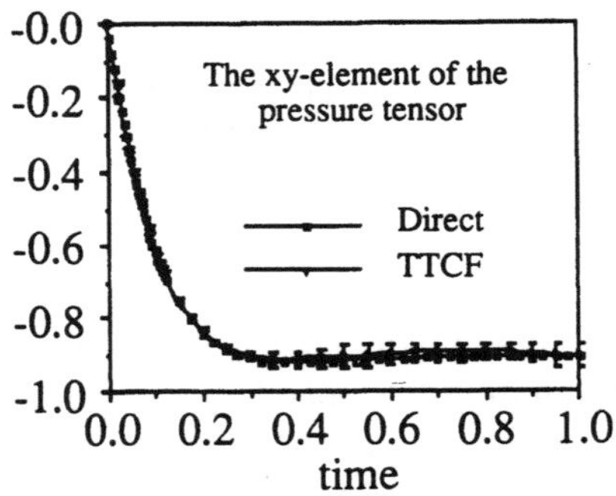

Figure 3. The response of the shear stress, $-P_{xy}$, in the soft disk fluid.

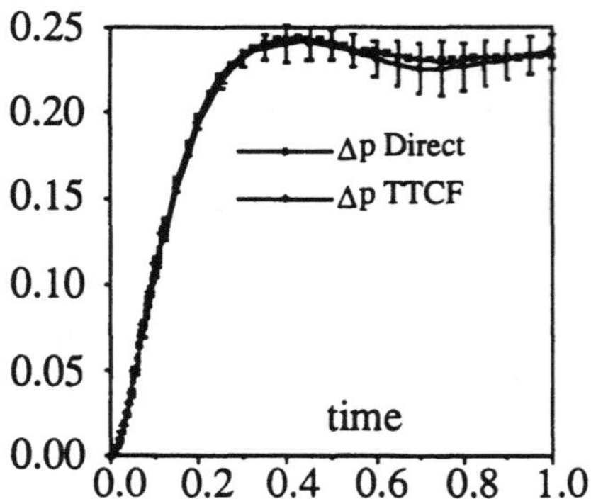

Figure 4. The response of hydrostatic pressure, $(P_{xx}+P_{yy})/2$ in the soft disk fluid.

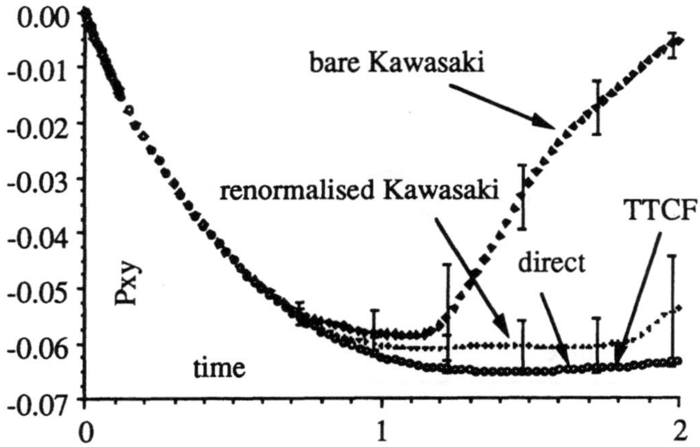

Figure 5. A comparison of direct, TTCF and Kawasaki responses for 18 soft disks undergoing shear flow, γ=1.

This prompted us to make a direct numerical check of the Kawasaki expression for the nonlinear response. In figure 5 below we compare the TTCF and Kawasaki averages for the nonlinear response of the shear stress [1]. The system consists of 18 soft disks at a reduced temperature and density of 1.0 and 0.1 respectively. The low density and small system size make the computations easier by reducing the magnitude of the Kawasaki exponent.

As can be seen from the figure the direct NEMD values for the xy-element of the pressure tensor, the TTCF values and the renormalised Kawasaki values, are in statistical agreement with each other. At long times the bare, or un-renormalised Kawasaki expression is apparently incorrect. These calculations are extremely difficult. The calculations shown in figure 5 comprise 8,000,000 starting states of 500 timesteps each (ie a total of 4 billion timesteps or 32 microseconds if the system is taken to represent argon!). Because of the extreme length of these simulation runs, they provided an accurate check of the accuracy of the TTCF formalism. In figure 5 the direct NEMD and TTCF results are so close to each other that they cannot be distinguished.

In figure 6 we show the relative deviation of the TTCF expression for the shear stress from the direct result. It can be seen that the maximum deviation is less than 2 parts in a thousand with a typical deviation being about one part in a thousand. This is the most accurate numerical check of either the Green-Kubo or the TTCF formalism that has yet been carried out.

3. Steady State Fluctuations

The Transient Time Correlation Functions and Kawasaki forms of the nonequilibrium distribution function may be considered as nonequilibrium

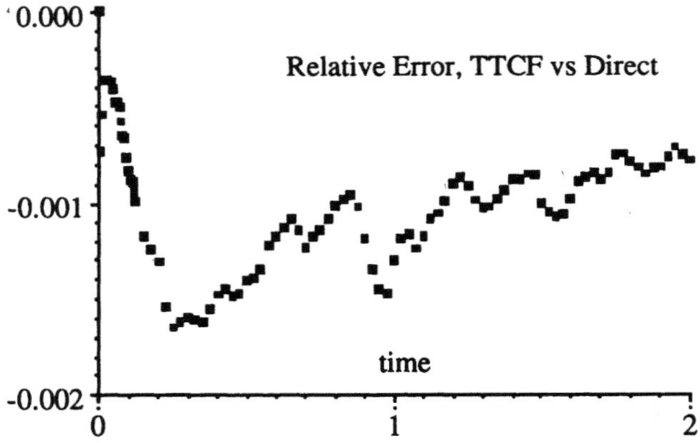

Figure 6. The relative deviations of the TTCF and direct NEMD for the soft disk fluid, γ=1.

partition functions. Formal expressions for nonequilibrium averages may be differentiated with respect to state-defining variables such as temperature, density or shear stress. In this way one can derive fluctuation formulae for properties of steady states such as specific heats and compressibilities [1]. For example an expression for the specific heat, $C_{V,\gamma}=\partial\langle H_0\rangle/\partial T_k|_{V,\gamma}$, can be derived from the renormalised Kawasaki expression for the steady state energy of a shearing system [1].

$$C_{V,\gamma} = k_B\beta^2\left\langle\Delta(H_0^2)\right\rangle_{ss} + k_B\beta^2\gamma\int_0^\infty ds\left\langle\Delta H_0(t)\Delta P_{xy}(s)V\right\rangle_{ss} . \qquad (4)$$

In this equation the ensemble averages, denoted $<\ ...\ >_{ss}$ are to be evaluated in the isokinetic steady state after the transients have decayed. One must use the renormalised rather than the 'bare' Kawasaki expression in deriving equation (4). For the Lennard-Jones ($\phi_{ij}=4\varepsilon[(\sigma/r)^{12}-(\sigma/r)^6]$), triple point system (T=0.722, n=0.8442), at γ=1, $C_{V,\gamma}/N$=2.37±0.06. The first term in the Kawasaki fluctuation expression for the specific heat is 3.31±0.02. The integral of the steady state stress-energy correlation function is -1.05±0.07 giving a total Kawasaki specific heat of 2.26±0.09 in statistical agreement with the result computed by numerical differentiation. For comparison the equilibrium (γ=0), specific heat is 2.662±0.004. Using similar methods fluctuation expressions can be derived for the isobaric specific heat, the compressibility [1] and for the nonlinear Burnett coefficients [1].

4. Lyapunov Instability and Response Theory

It is well known that the equations of motion for statistical mechanical systems are subject to Lyapunov instability [6]. An initial displacement in phase space

$\delta\Gamma(0)$ at t=0 increases exponentially in time at a rate which is equal to the largest Lyapunov exponent λ_{max}. Throughout the following we use constant internal energy, H_0, dynamics. Consider the equations of motion for a set of orthogonal tangent vectors, $\{\delta\Gamma_i(t);\ i=1,...,6N\}$, describing the progressive separation of two phase trajectories initially separated by $\delta\Gamma_i(0)$. The set forms a complete orthogonal basis for the tangent space with $\delta\Gamma_1(t)$ being parallel to the direction of fastest growing phase separation, $\delta\Gamma_2(t)$ being orthogonal to $\delta\Gamma_1(t)$ and $\delta\Gamma_1(t)\times\delta\Gamma_2(t)$ generating the fastest growing area, ... etc [6]. The equations of motion for the infinitesimal tangent vectors are

$$\frac{d}{dt}\delta\Gamma_i(t) \equiv \mathbf{T}(\Gamma)\bullet\delta\Gamma_i(t) = \frac{\partial\dot{\Gamma}(\Gamma(t))}{\partial\Gamma}\bullet\delta\Gamma_i(t), \qquad (i=1,..,6N). \tag{5}$$

In the infinitesimal limit, $\delta\Gamma_i(0)\rightarrow 0$, the formal solution of this equation can be written as [6], $\delta\Gamma_i(t) = \mathbf{L}(t)\bullet\delta\Gamma_i(0)$. The Lyapunov exponents can be defined as the logarithms of the eigenvalues of the symmetric matrix, $\mathbf{\Lambda}$, [6]

$$\mathbf{\Lambda} = \lim(t\rightarrow\infty)\mathbf{\Lambda}(t) = \lim(t\rightarrow\infty)\left[\mathbf{L}^T(t)\bullet\mathbf{L}(t)\right]^{1/2t}. \tag{6}$$

Until recently there was no known connection between the Lyapunov exponents and thermophysical properties such as the shear viscosity. The Liouville equation [7] states that, $(1/f)df/dt=3N\alpha$. We can see that the accessible volume of phase space, $V\sim 1/f$, decreases to zero

$$\langle d\ln V/dt\rangle = -3N\langle\alpha\rangle = -\Sigma_1^N\lambda_i\ . \tag{7}$$

This happens because the *dimension* of the steady state phase space decreases to less than the ostensible 6N Cartesian dimensions. We call equation (7) the Lyapunov Sum Rule.

The 6N Lyapunov exponents can be ordered in terms of their magnitude. One defines conjugate pairs of exponents by grouping the largest with the smallest, the second largest with the second smallest, .. etc. It is well known that for Hamiltonian systems the sum of each conjugate pair, known as Smale pairs [6], $\lambda_i+\lambda_{i'}$, is zero. Very recently we have proved [8] that for Gaussian thermostatted Hamiltonian systems, the Lyapunov exponents obey the Conjugate Pairing Rule,

$$\lambda_i + \lambda_{i'} = -\langle\alpha\rangle = 2\bar{\lambda}\ . \tag{8}$$

This rule is also true for thermostatted shearing steady states, described by the SLLOD equations of figure 2.

Equation (8) is of considerable practical and theoretical value. The largest Lyapunov exponent is the easiest to evaluate followed by the second largest etc. The most negative exponent is normally **very** difficult to calculate [6]. However the most negative Lyapunov exponent for a steady state is the negative

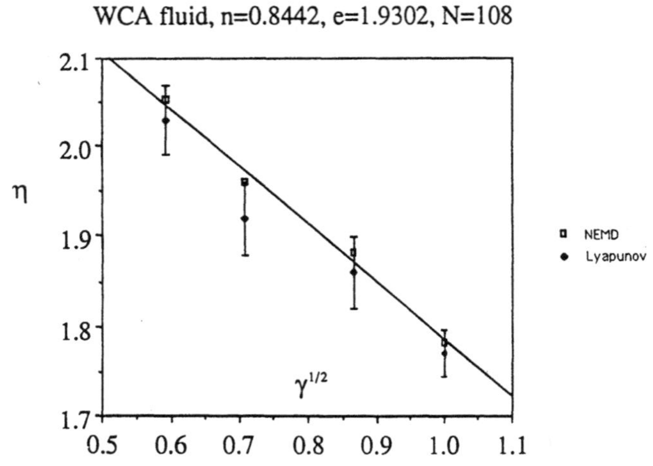

Figure 7. We compare the shear viscosities for the WCA fluid $\phi_{ij}=4\varepsilon[(\sigma/r)^{12}-(\sigma/r)^6]+1$, computed using NEMD and the Conjugate Pairing Rule. Both sets of data are in agreement and are consistent with a linear dependence on $\gamma^{1/2}$, indicated by the straight line.

of the largest exponent for the *antisteady* state - that state formed from the steady state by reversing the direction of time. This observation also allows λ_{min} to be evaluated with ease. One must of course determine this exponent in a time less than the decay time of the antisteady state, $t_{1/2}$, otherwise one would obtain the same exponent as for the steady state (see figure 8).

Combining equations (7) and (8) one can derive an exact relation between the sum of the two *maximal* Lyapunov exponents and the shear viscosity. Exploiting the relation of the SLLOD equations of motion, to symplectic mappings one can show that [8]

$$\eta(\gamma) = -3nk_B T[\lambda_{max}(\gamma) + \lambda_{min}(\gamma)] / \gamma^2. \tag{9}$$

We exhibit the results of computer simulations utilising equation (9) in figure 7.

5. Conclusions

There are many measures of the dimension of accessible phase space. Two of these measures, the Lyapunov dimension and the information dimension [1,9], have been computed in computer simulations. It is found that as one moves further from equilibrium, the accessible dimension decreases and, in the long time limit, the N-particle distribution becomes fractal [1,10]. This is true in both the linear and nonlinear response regimes.

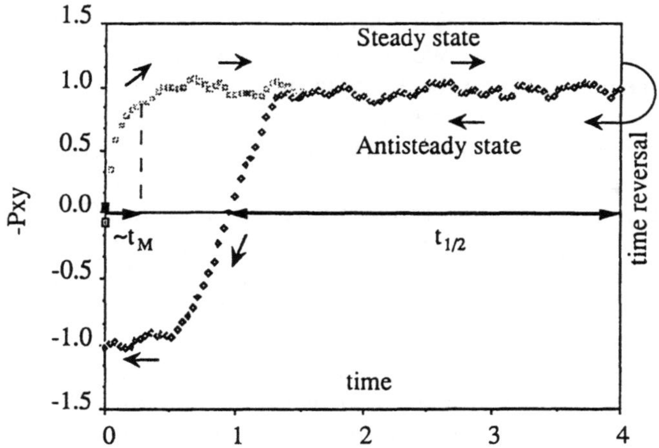

Figure 8. We show the shear stress, $-P_{xy}$, as a function of time for the steady and antisteady states. The system is: 108 WCA particles at: $\gamma = 0.5$, n=0.8442, E/N=1.9302. The antisteady state is unstable and eventually decays after a time $t_{1/2}$, generating the steady state for the reversing system. The lifetime of the anti-steady state, $t_{1/2}$, is significantly greater than the time required to relax from equilibrium into the steady state $\sim 2t_M$.

It may seem that the fractal nature of the nonequilibrium N-particle distribution function might call into question [11] the usual mathematical machinery used in the formal solutions of the Liouville equation. Such solutions form the basis of Green-Kubo, the TTCF and the Kawasaki expressions for nonequilibrium averages. These solutions assume that the relevant distributions in both time and space, are sufficiently smooth that formal power series expansions can be used to represent the time dependent solutions of the Liouville equation. One can demonstrate that in spite of the exponential divergence of nearby phase space trajectories, the irreversibility generated by Lyapunov instability and the associated fractal nature of the distribution function manifest themselves sufficiently slowly that microscopic trajectories can be *reversibly* followed for times long enough that the relevant correlation integrals (1 - 3) converge. Thus although this objection is valid in principle, it is not relevant in practice.

In figure 8 below we show the response of the shear stress to a suddenly imposed strain rate. The system initially evolves in positive time. We display $P_{xy}(t)$ for an individual trajectory . Within a couple of Maxwell times, t_M, the xy-element of the pressure tensor relaxes to its steady state value. Once t = 4.0, the direction of time is reversed. The reversed trajectory follows the forward trajectory very closely until Lyapunov amplified noise becomes comparable to typical particle velocities or separations. Until this time, the system has exhibited a **negative** viscosity and a negative rate of entropy production. The thermostat is supplying heat to the system which is converted into work and drives the shear. This of course violates the Second Law of Thermodynamics. We call this state the *antisteady state*. We define the lifetime of the antisteady state, $t_{1/2}$, in terms of the time taken for the decaying stress to cross zero.

The equations of motion were solved using a 4th order Runge-Kutta algorithm with a typical reduced timestep for simulations of 0.005. Relaxation to the steady state takes place in about $2t_M$~0.5 but $t_{1/2}$~3.0 ~ $6t_M$. This means, for times less than $t_{1/2}$ the Lyapunov instability is too small to affect the computed trajectories. Propagators and their inverses can be reliably modelled by the Liouville equation used in response theory. For response theory to be valid it is only required that formal operator manipulations are correct for times needed for relaxation to the steady state ~$2t_M$.

Thus we see that the objection to response theory [11] is not relevant because the objection only applies at times *after which* the relevant TTCF or Kawasaki integrals have converged. The objection is closely related to the van Kampen objection [12] to linear response theory and is dismissed [13] for the same reasons.

If a computer simulation is inaccurate and the antisteady state lifetime is shorter than the time required by the system to relax into the steady state, then not only are the TTCF and Kawasaki formalisms likely to lead to results which differ from direct NEMD, but the direct NEMD results themselves are also likely to be incorrect. Simulations should be performed with sufficient accuracy for trajectories to be reversible over the timescale required for convergence of the relevant TTCF and Kawasaki integrals.

The Lyapunov Sum Rule, (7), and the Conjugate Pairing Rule, (8,9), show that Lyapunov instability and chaotic dynamics are relevant for the understanding of nonequilibrium processes and the calculation of transport coefficients. In fact the existence of Lyapunov exponents in the contracting phase space of thermostatted dissipative systems is intimately related to the very existence of transport coefficients.

Acknowledgement

We would like to thank the Australian National University Supercomputer Facility for a generous grant of computer time.

References

1 D J Evans & G P Morriss *Statistical Mechanics of Nonequilibrium Liquids*, Academic Press, London (1990).
2 G P Morriss & D J Evans, Mol. Phys., **54**, 629 (1985).
3 T Yamada and K Kawasaki, Prog. Theo. Phys., **38**, 1031 (1967).
4 R Kubo, J. Phys. Soc. Jpn., **12**, 570 (1957),
M.S. Green, J. Chem. Phys., **22**, 398 (1954).
5 Q W Lees & S F Edwards, J. Phys. C., **5**, 1921 (1972).

6 J -P Eckmann and D Ruelle, Rev. Mod. Phys., **57**, 617 (1985).
7 W.G. Hoover and H.A. Posch, Phys. Letts., **123A**, 227 (1987).
8 D J Evans, E G D Cohen and G P Morriss, Phys. Rev. A, (to appear).
9 H G Schuster, *Deterministic Chaos*, VCH Verlagsgesellschaft, Weinheim (1989).
10 G P Morriss, Phys. Letts., **A122**, 236 (1987), **A143**, 307 (1989).
G P Morriss, Phys. Rev., **A37**, 2118 (1988), **A39**, 4811 (1989).
H A Posch and W G Hoover, Phys. Rev., **A38**, 473 (1988), **A39**, 2175 (1989).
11 B L Holian, G Ciccotti, W G Hoover, B Moran and H A Posch, Phys. Rev., **A39**, 5414 (1989).
12 N G van Kampen, Acta Phys. Norvegica **5**, 279 (1971).
13 G P Morriss, D J Evans, E G D Cohen and H van Beijeren, Phys. Rev. Letts., **62**, 1579 (1989).

Two-Phase Coexistence of String and Liquid Phases in Planar Couette Flow

S. Nosé and T. Yamada

Department of Physics, Faculty of Science and Technology,
Keio University, 3-14-1 Hiyoshi, Kohokuku, Yokohama 223, Japan

Abstract. The orientation of the two-phase boundary between the string and liquid phases in the planar Couette flow is investigated. A specific orientation has always been obtained in former simulation studies. We studied why this orientation is stable. We start simulations from configurations obtained by rotating the stable two phase coexistence configurations 90° around the flow direction. A change from the rotated to the stable orientations is observed in one of our simulations.

1. Introduction

The nonequilibrium molecular dynamics (NEMD) simulation method is developed to obtain an accurate value of transport coefficients efficiently. The viscosity of a fluid is determined as a response to a perturbation which causes a velocity gradient in a flow. To obtain the viscosity, the planar Couette flow which has a linear velocity profile $u_x = \gamma y$ is produced by a nonequilibrium molecular dynamics method. u_x is the flow velocity. We define the x axis as the flow direction and y as the velocity gradient direction. The viscosity η is determined as a ratio of the xy component of the shear stress tensor P_{xy} and the velocity gradient $\gamma = \partial u_x / \partial y$, $\eta = -P_{xy}/\gamma$. The viscosity in equilibrium is obtained as the value of η in the limit of $\gamma \to 0$.

In 1984, Erpenbeck found a new peculiar phase, which is now known as the string phase, in simulations in the large shear rate region [1]. The particles move parallel to the flow, aligning into strings, and these strings form a triangular lattice in a plane perpendicular to the flow. Many nonequilibrium simulations have been carried out encouraged by the discovery of the string phase [2-6].

A thermostat is always used in this type of simulation to take away the heat produced by friction. Only the deviation of the velocity from the assumed linear velocity profile was considered to contribute to the temperature. Evans and Morriss [5] pointed out that the assumption of a linear velocity profile is not appropriate at a large shear rate region and that the use of a thermostat which does not assume a linear velocity profile suppresses the appearance of the string phase.

They separated a MD cell into blocks, and the flow in a block is determined by averaging the velocities of particles in the block. The temperature is related to the fluctuation of the velocity around the averaged flow thus obtained. However, in their simulation, the averaged number of particles in a block was only two. Loose and Hess [6] pointed out that the average velocity in the Evans-Morriss work is not appropriate one. Loose and Hess separated a MD cell into layers along the y direction to get good statistics on averaging. They observed a change relating to an instability of a linear velocity profile and obtained a phase similar to the string phase in a large velocity gradient region.

We are interested in the orientation of the boundary between the string and liquid phases in a two phase coexistence region which appears in an intermediate velocity gradient region. The boundaries are oriented parallel to a xy plane in every NEMD simulation assuming a linear velocity gradient profile as far as we know [1-4]. Fig.1 shows a configuration in the two phase coexistence region. The centers of mass of all the particles in the system are projected onto a yz plane. The regions forming a triangular lattice are in the string phase. A liquid phase in the middle region shows no significant ordering. We are interested in why this orientation is always chosen and what is the determining factor to stabilize this orientation. We start our study from the investigation of whether the orientation found in the simulations is really stable or not.

Configurations obtained by simulations in the two phase coexistence region are rotated 90° around the x axis. Our simulations are started from these rotated configurations. We obtained the result that the internal energy and the viscosity are larger in the rotated configuration than in the original configuration. This result confirms that the configuration usually obtained in simulations is really more stable than the rotated one. This agrees well with former simulations. We observed a change from a 90° rotated configuration to the stable one in a simulation with the velocity gradient near the transition point from the liquid to the string phase. This is a clear confirmation of our conclusion.

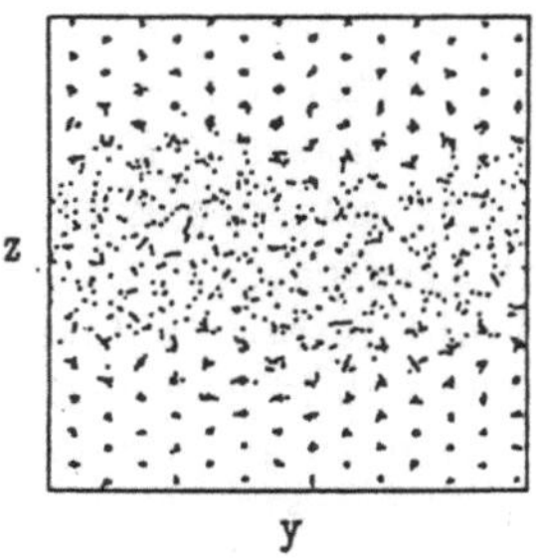

Figure 1
A view from the flow direction of two phase coexistence. The y axis is parallel to the velocity gradient. 2048 Lennard-Jones system. γ=3.00,T=0.71, and ρ=0.8174.

We also studied the distribution of several quantities along the velocity gradient direction.

2. Nonequilibrium simulation method

The Lees-Edwards sliding MD cell periodic boundary condition [7] and the SLLOD algorithm [8] are employed to produce the Couette flow with a linear velocity profile. The Lees-Edwards condition is devised to conform to the external condition with the velocity gradient and the periodic boundary conditions. A MD cell slides along the x direction with the relative velocity γL on a MD cell directly below. L is the height of a MD cell along the y direction. Thus, the periodic conditions for the x component of the coordinate and the velocity in the y direction become

$$x(y+L) = x(y) + \gamma Lt \ , \tag{1}$$

$$v_x(y+L) = v_x(y) + \gamma L \ , \tag{2}$$

where t is the time from the start of the simulation. Ordinary periodic boundary conditions are applied in the x and z directions.

In the SLLOD algorithm, the following modified equations of motion are used:

$$\begin{aligned}
\frac{dx_i}{dt} &= \frac{p_{ix}}{m_i} + \gamma y_i \ , & \frac{dp_{ix}}{dt} &= -\frac{\partial \Phi}{\partial x_i} - \gamma p_{iy} - \zeta p_{ix} \ , \\
\frac{dy_i}{dt} &= \frac{p_{iy}}{m_i} \ , & \frac{dp_{iy}}{dt} &= -\frac{\partial \Phi}{\partial y_i} - \zeta p_{iy} \ , \\
\frac{dz_i}{dt} &= \frac{p_{iz}}{m_i} \ , & \frac{dp_{iz}}{dt} &= -\frac{\partial \Phi}{\partial z_i} - \zeta p_{iz} \ .
\end{aligned} \tag{3}$$

The new terms with a coefficient γ work to produce a velocity gradient. The terms with a coefficient ζ appear to keep the temperature at a constant value. In the above equation, the velocity dx_i/dt consists of two terms; γy_i, the flow velocity, and p_{ix}/m_i, a thermal fluctuation around the flow velocity. The assumption of the linear velocity profile is clearly expressed in this equation. A term $-\gamma p_{iy}$ expresses a friction force which is caused by particle motion in the y direction and works to reduce the velocity gradient.

The heat produced by friction between sliding fluid layers is taken away by a Gaussian thermostat [9], which is expressed in the terms with a variable ζ in equation (3). The value of the total kinetic energy $K=\Sigma_i(p_{ix}^2+p_{iy}^2+p_{iz}^2)/2m_i$ is kept constant. The coefficient ζ

is determined by differentiating the constraint of the constant kinetic energy.

$$\Sigma_i (p_{ix}\dot{p}_{ix} + p_{iy}\dot{p}_{iy} + p_{iz}\dot{p}_{iz}) = 0 . \tag{4}$$

Thus, ζ is

$$\zeta = -(\Sigma_i p_i \cdot \frac{\partial \Phi}{\partial r_i} + \gamma \Sigma_i p_{ix} p_{iy})/(\Sigma_i p_i^2) \tag{5}$$

We studied 256 and 2048 particle systems which interact via a Lennard-Jones 12-6 potential, and all units are reduced in units of m (mass of a particle), ε (energy, well depth of a potential), and σ (length, core size). The potential is truncated at 2.9. Most calculations are carried out in the liquid phase at temperature T=0.71 and density ρ=0.8174. This condition is near the triple point of a Lennard-Jones system. The time increment is chosen as Δt=0.001 for shear rate $\gamma = \partial v_x / \partial y$=2.42 and 3.00 in which we did long simulations with 2048 particles.

3. Results

The viscosity of the Couette flow is calculated using the formula $\eta = -\langle P_{xy} \rangle / \gamma$. The viscosity decreases linearly with respect to $\sqrt{\gamma}$ in the liquid phase. At γ>2.4, the viscosity decreases suddenly. This phenomenon is known as shear thinning. The string phase appears with this sudden decrease of viscosity. At first, it coexists with the liquid phase. The whole system becomes the string phase in the large shear rate region.

The orientation of the two phase boundary is parallel to the xy plane, which agrees with former simulations. A primitive lattice vector of the triangular lattice is parallel to the z axis. New configurations are produced by rotating configurations obtained by simulations 90° around the x axis. A primitive lattice vector points to the y axis in the rotated configuration. The two phase boundary orients perpendicular to the y axis. Longer simulations starting from the rotated configurations were carried out at γ=2.42 and 3.0 with 2048 particles.

At γ=3.0, the orientation and structure of the two-phase coexistence was stable during our simulation of 60000 steps. There was no indication of structure change during this calculation. The total energy per particle E/N increased by 0.35 from -5.03 to -4.68, and the viscosity increased by 0.28 from 0.48 to 0.76 within less than 2000 time steps after the rotation. A larger viscosity means more work is necessary to maintain a velocity gradient. Thus, the increase of the

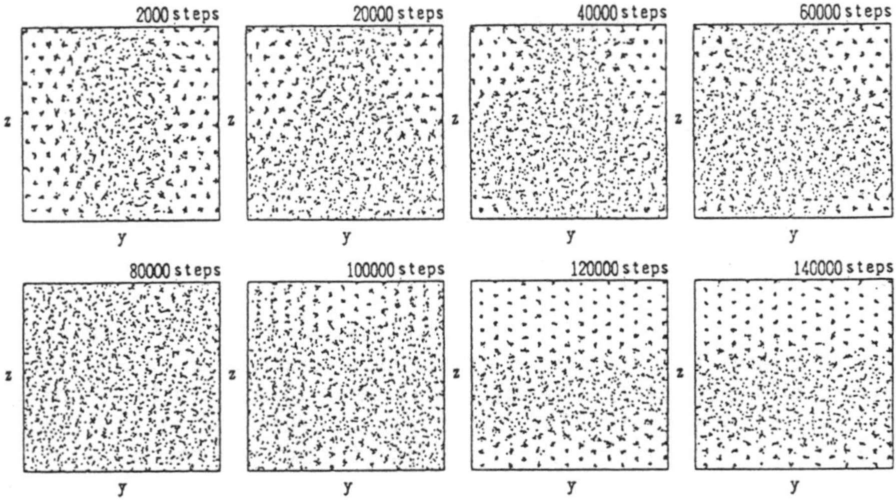

Figure 2
Change of the orientation of two phase boundary at γ=2.42, T=0.71, ρ =0.8174, and N=2048. Time steps after the rotation are indicated at the top right.

total energy and the viscosity indicate that the orientation usually obtained by simulations is really more stable than the rotated one.

At γ=2.42, which is very close to the transition point from the liquid to the two phase coexistence, the rotated structure becomes unstable and we observe a change from a 90° rotated configuration to the stable orientation perpendicular to the z axis.

The temporal evolution of the atomic configuration in this process is depicted in Fig.2. At 2000 steps after the rotation, the boundary still remains in the orientation perpendicular to the y axis. At 20000 steps, already a part of the string phase changes to the liquid phase.The string phase is no longer extended to the size of a MD cell, and has a circular boundary surrounded by the liquid phase. At 60000 steps, the string phase region is diminishing, and at 80000 steps, there is no indication of the string phase. The whole system appears to be in a liquid state. A new string phase ordering in which a primitive lattice vector points to the z axis appears a little later, extends along the y direction, and finally forms a stable two-phase boundary perpendicular to the z axis.

The time development of the total energy E/N before and after the rotation is shown in Fig.3. The change of the viscosity is similar to that of the total energy: Just after the rotation, the total energy increases quickly from -4.90 to -4.65, the same as for γ=3.0. The rotated configuration is unstable in comparison with the original one.

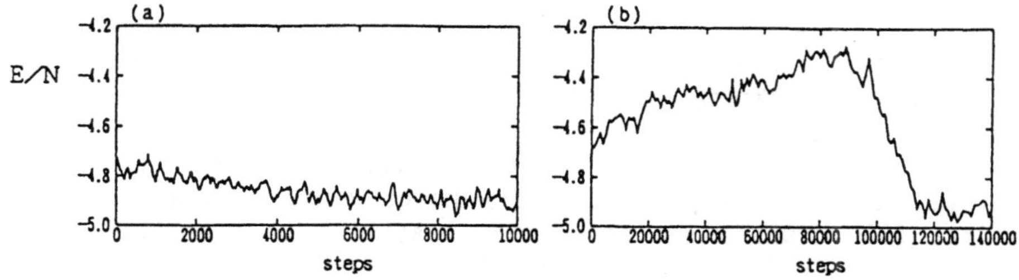

Figure 3
The time evolution of the total energy E/N a) before and b) after the 90° rotation. The condition is the same as in Fig.2.

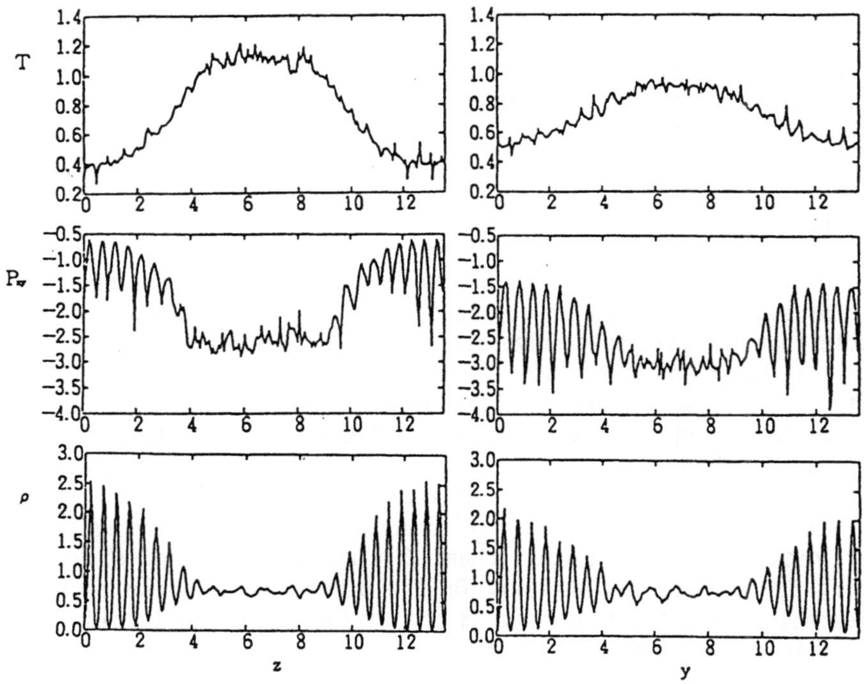

Figure 4
The distribution of the temperature, the xy component of the stress tensor, and the density. The left column is in the stable orientation, and the right column is in the rotated orientation. The condition is the same as in Fig.2.

The total energy increases as the area of the string phase decreases. The energy reaches a maximum value at about 80000 steps, where the ordering of the string phase almost disappears. Thereafter, the energy starts to decrease rapidly to the same value before the 90° rotation.

The distributions of the temperature, density and the xy component of the stress tensor P_{xy} along the y direction are calculated in the first 5000 times steps after the rotation. The y direction is separated into 200 layers. In the calculation of the stress tensor, a virial term relating to two different layers is divided equally between the two layers. In this way, the average of the stress tensor obtained from the distribution coincides with the average in the whole system.

In Fig.4, the distributions are depicted both in the stable orientation and in the rotated orientation at γ=2.42. The temperature is high and the shear stress is low in the liquid region. The density distribution in the string phase region clearly shows its crystal-like ordering. The difference of the energy and the shear stress between the liquid and the string phases is larger in the stable orientation than in the rotated one.

4. Discussion

As shown in the preceding section, we confirmed that the orientation of the two phase boundary obtained in former simulations is more stable than the rotated configuration. What is the determining factor in the stability of shear induced structures?

We consider a phenomenological model. A system is separated in blocks. The energy dissipation occurs mainly by the friction between the sliding blocks. A structure with smaller energy dissipation will be stabilized. In the stable orientation, only the frictions between the same phases (λ_{SS} and λ_{LL}) contribute. $\lambda_{\alpha\beta}$ is the friction coefficient between two blocks α and β (subscripts α and β denote string S, or liquid L phases). On the other hand, in the rotated orientation, the friction between the string phase and liquid λ_{SL} also arises. The sign of $2\lambda_{SL} - \lambda_{LL} - \lambda_{SS}$ determines the relative stability of the two orientations. Our conclusion that the original orientation is stable means $2\lambda_{SL} > \lambda_{LL} + \lambda_{SS}$. In the string phase, the friction is supposed to be considerably smaller than in liquid $\lambda_{LL} \gg \lambda_{SS}$. The determining factor will be reduced to how large λ_{SL} is in comparison with λ_{LL}.

The distribution of the stress tensor along the y direction confirms microscopically the relation $\lambda_{LL} \gg \lambda_{SS}$. The difference of the temperature in the liquid and the string phase regions reflects the difference of the heat production (thus, the difference of the friction) in two regions. The temperature is only controlled uniformly in the whole system. A thermostat does not work to smooth the temperature distribution. The transition in the boundary region from the liquid to the string phase is continuous. Therefore, it is not easy, at present, to determine the friction at the boundary between two phases and to confirm the result of a phenomenological model.

We remark on another insufficient aspect of our phenomenological model. The temperature difference between the string and the liquid phases is different in the two configurations. The difference is smaller in the rotated orientation. This means that the heat production in the string phase is higher in the rotated orientation than in the stable one. Therefore, the stability of the string phase is dependent upon the orientation of the triangular lattice to the velocity gradient direction. This factor is not taken into consideration in our phenomenological model. We are now continuing our investigation to separate the effects in the orientation of the triangular lattice and in the orientation of the two phase boundaries.

References

[1] J.J.Erpenbeck, Phys.Rev.Lett. 52, 1333 (1984)
[2] L.V.Woodcock, Phys.Rev.Lett. 54, 1513 (1985)
[3] D.M.Heyes, J.Chem.Soc.Faraday Trans.2, 82, 1365 (1986)
[4] D.M.Heyes, J.Chem.Phys. 85, 997 (1986)
[5] D.J.Evans and G.P.Morriss, Phys.Rev.Lett. 56, 2173 (1986)
[6] W.Loose and S.Hess, Rheol. Acta 28 91 (1989)
[7] A.W.Lees and S.F.Edwards, J.Phys.C 5, 1921 (1972)
[8] D.J.Evans and G.P.Morriss, Phys.Rev.A 30, 1528 (1984)
[9] D.J.Evans,W.G.Hoover,B.H.Failor,B.Moran and A.J.C.Ladd, Phys.Rev.A 28, 1016 (1983)

Part VII

Interfaces and Ionic Conductors

Simulation Studies of Interfacial Phenomena – Melting, Stress Relaxation and Fracture

S. Yip

Department of Nuclear Engineering, Massachusetts Institute of Technology, Cambridge, MA 02139, USA

Abstract. Atomistic simulation offers unique capabilities for probing the complex molecular interactions at materials interfaces, processes which are fundamental to our understanding of materials properties and behavior. We describe three recent studies which have provided insights into the role of interfaces in crystal melting, elastic interactions on stepped surfaces, and brittle-ductile transition in crystal fracture.

1. Introduction

The capability of atomistic simulation methods such as molecular dynamics and Monte Carlo in modeling complex physical behavior is by now well recognized in condensed matter physics, statistical mechanics, and theoretical chemistry. However, their applications to materials science have not been sufficiently extensive or definitive to the extent that the advantages and limitations of these techniques are well demonstrated. The purpose of this contribution is to discuss several recent results on interfacial phenomena; collectively they serve to illustrate the considerable opportunities that exist for simulation studies to contribute to an important area of materials research.[1]

2. Thermodynamic and Mechanical Melting

Despite the extensive efforts in studying the phenomenon of melting [2], there are still aspects of this fundamental phase transition which have not been clarified. One basic question which has been raised recently [3] is the role of surfaces or interfaces in the mechanism of melting. From the standpoint of thermodynamics, melting occurs at the temperature at which the solid and liquid phases coexist, as expressed by the equality of the Gibbs free energies. However, thermodynamics says nothing about how melting occurs or how long the process will take; these are issues pertaining to the kinetics of the phenomenon. Thus one can ask whether our thermodynamic picture of melting is one which is consistent with the kinetics.[4] This is a question which can be addressed by molecular dynamics in the sense that simulation provides a method to calculate the free energies of the solid and liquid phases, as well as to directly observe the actual melting process at the molecular level.[4]

In any simulation study related to melting it is essential to recognize that the melting point of the simulation model, T_m, can be quite different from the known melting point of the real substance. How well these two temperatures agree is, in fact, a useful indication of how realistic is the interatomic potential function associated with the model. We will examine the question of the interplay between thermodynamics and kinetics in the particular case of a simulation model of silicon based on the empirical potential function developed by Stillinger and Weber.[5] For this potential free energy calculations have been carried out by Broughton and Li which give the melting point T_m at 1691 ± 20 K.[6] It should be noted, in passing, that the melting point of silicon is 1683 K, and that the remarkable agreement should be regarded as somewhat fortuitious.

Once T_m is determined, one then has the proper reference temperature from which to investigate the onset of melting. By taking a perfect crystal model of silicon, composed of 704 atoms in a cell with periodic boundary conditions, and heating it up to T_m using molecular dynamics, it is found that over a reasonable period of simulation the system shows no indications of any onset of structural disordering.[4] This apparent stability is found to persist up to temperatures well beyond T_m; it is only when T reaches 2500 K that the system is observed to become significantly disordered over a period of about 16000 time steps (about 0.18 ps). Thes observations are perplexing at first sight. Why did the simulation model not melt at T_m as predicted by thermodynamics? Should the disordering at 2500 K, hereafter denoted as T_s, be interpreted as the onset of melting?

To answer these questions another set of simulations is performed using a system cell which has free surfaces in one direction and periodic borders in the other two directions.[4] These runs are made at temperatures above T_m but below T_s. It is observed that structural disordering, which may be regarded as local melting, begins invariably at the surface and then spreads toward the interior of the cell. For a quantitative measure of local disorder we divide the cell into 32 equal slices in the direction of the surface normal, and calculate the static structure factor $S(\underline{K})$, with $\underline{K}$ chosen to be a reciprocal lattice vector with orientation parallel to the surface, for each individual slice. From the profiles of $S(\underline{K})$ obtained at various intervals during the simulation it is possible to locate the melt-crystal interface, and by following this interface in time one can determine its speed of propagation, v(T). Similar studies have been carried out for a model with an internal surface, a grain boundary.[4]

The data on the propagation speed thus determined are shown in Fig. 1. By fitting the data with a quadratic curve one finds that v(T) extrapolates to zero at a temperature of 1710 ± 30 K.

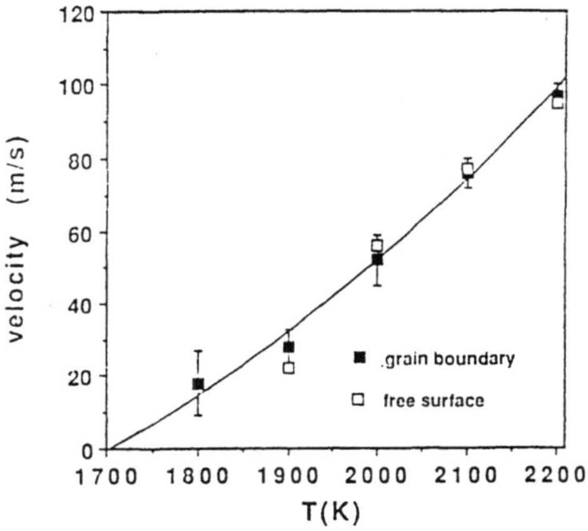

Fig. 1. Propagation speed of melt-crystal interface in silicon obtained from melting induced by the presence of a grain boundary or a free surface.[4]

Since the temperature at which the interfacial velocity vanishes is that temperature at which the melt and the crystal are in equilibrium, this is then another way of determining the melting point of the system. We see that within the uncertainties of the numerical results the extrapolated temperature agrees with T_m. The same conclusion has been reached in a parallel study using an appropriate potential for Cu.[4]

Our findings therefore show that thermodynamics is indeed consistent with kinetics; both methods of determining the melting point of the system give the same value of T_m. The observation that with periodic borders in all directions the system is stable up to a temperature T_s means that the system can be easily superheated in the absence of an interface. This is evidence that there exist two kinds of melting phenomena, a heterogeneous process of nucleation and growth at a surface or interface, and a homogeneous process of mechanical instability against shear. The former is what we can call thermodynamic melting; it occurs at T_m where the Gibbs free energies of the solid and liquid phases are equal. The latter is sometimes known as <u>mechanical</u> melting; it occurs at a temperature T_s where the elastic constants C_{11} is equal to C_{12} and the system is unstable against shear deformation.[4]

3. Stress Relaxation on Stepped Surfaces of Silicon (100)

In the initial stage of epitaxial growth of semiconductor materials, the perfection of the substrate surface is particularly critical since it is the template for the growth of

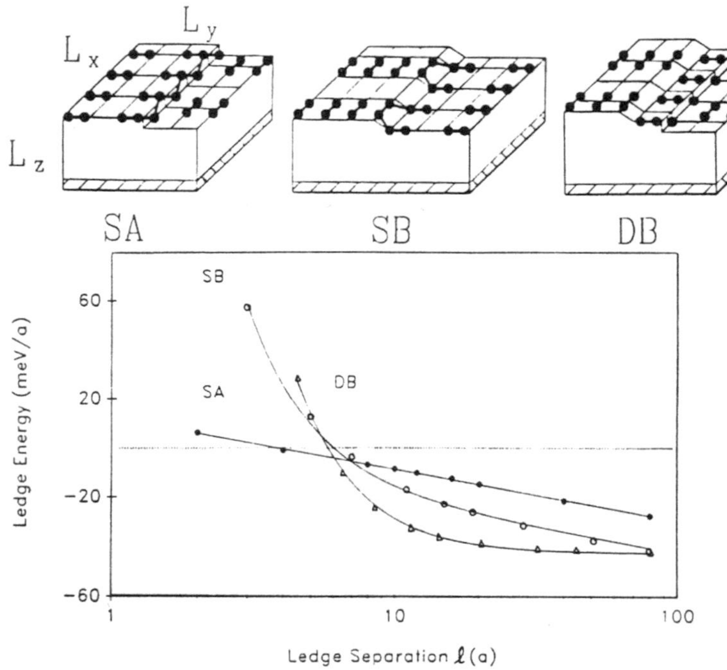

Fig. 2. Structural models and ledge energies of stepped surfaces on Si(100).

the overlayers. Recently experimental observations of stepped Si(100) using LEED and STM have shown the existence of both single- and double-layer ledges and therefore provided the necessary data to check the theoretical predictions of relative stability of these stepped surfaces.[7] The structural models which have been analyzed in electronic-structure calculations are two single-layer structures SA and SB, and a double-layer structure DB.[8] As illustrated in Fig. 2, while the 2x1 reconstruction occurs on all surfaces, the A-type structure is distinguished by the dimer axis on the upper terrace being perpendicular to the ledge, while in the B-type structure the dimer is parallel to the ledge. For a single-layer step the dimers on the upper and lower terraces must alternate in their orientations as a result of the tetrahedral bonding. In contrast, for a double-layer step the dimers must have the same orientations.

The significance of the these structural models is that calculations based on tight-binding approximations have shown that for a relatively small system (short ledge separation in a periodic system) the structure SA is lower in energy than SB, and that DB is lower in energy than a cascade of SA + SB.[8] When one takes into account the variation of energy with ledge separation, it has been pointed out recently that stress relaxation can lower the energy of a stepped surface to below that of a flat surface.[9,10] By applying such an analysis to vicinal surfaces, it was found that a transition from a cascade of single-layer structures SA + SB to the double-layer structure DB occurs at a certain miscut angle θ_c.[11-13] This prediction can be tested against experimental observations as we will discuss below.

Static relaxation and molecular dynamics studies using the Stillinger-Weber potential function have been carried out on the structural models just discussed.[12,13] In these calculations the main interest is in the effects of ledge interactions and the determination of relative stability of single- and double-layer stepped structures. The simulation cells are periodic along the x direction, with dimension L_x equal to 4 atomic layers; it is also periodic along the y-direction with L_y being a variable. In the z direction the atoms on four bottom layers are held fixed and dimension L_z is taken to be large enough that bulk condition is obtained in the interior. For the systems reported here L_y varies from 6a to 102a, with a being the surface lattice constant (equal to 3.84 A), L_z varies from 22 to 134 layers, and the number of atoms in the simulation cell varying from 492 to 54,468.

We consider here only the results of energy minimization obtained by the conjugate gradient method. We define the ledge energy as $\lambda(\ell) = L_\ell^{-1} (E - N\epsilon_b - L_xL_y\epsilon_u)$, where E is the total energy, N the number of atoms in the simulation cell, ϵ_b the energy of a bulk atom (-4.33 eV for the SW potential), ϵ_u the surface energy of the uniform reconstructed 2x1 surface (eV/a^2), and L_ℓ is the length of the ledge ($2L_x$ for flat surface and L_x for vicinal surface). From the relaxed structures of SA, SB, and DB at various ledge separations $\ell = L_y/2$, we have evaluated $\lambda(\ell)$. The results are given in Fig. 2. One sees that all the stepped structures have energies lower than that of the 2x1 flat surface beyond a ledge separation of ~ 7a, thus indicating that energetically speaking the stepped structures of certain sizes are favored over the flat surface.
This crossover occurs first for SA, then DB, and lastly SB. We note that at small ℓ, SA is the lowest in energy among the three structures, while SB is the lowest at large ℓ.

In the case of SA we see that over the two decades of ℓ values, the simulation results show a logarithmic variation. A general argument based on elasticity theory has been given by Marchenko which shows that when two phases of different surface stress exist, the deformation (strain-field) energy varies logarithmically with the dimension separating the two phases [9]. This argument is applicable to the single-layer stepped structures since the upper and lower terraces are in different stress states by virtue of the dimerization orientations.

Indeed, by fitting our data on SA to the form $\lambda_o - \lambda_\sigma \ln(\ell/\pi a)$, we obtain $\lambda_\sigma = 9.07$ meV/a. This is in reasonable agreement with a value of 11.5 meV/a deduced from electronic-structure calculations using the tight-binding approximation [10].

In the case of DB, we observe a relatively strong repulsive ledge-ledge interaction at small ledge separations. It has been shown using linear elasticity that the force-dipole interaction between adjacent ledges gives rise to an inverse power law variation with an exponent equal to 2.[14] We therefore anticipate that an interaction of this type will be present in the DB structure because of the ledge-rebonding process mentioned above. Consequently, we find that λ(DB) can be well fitted to the form $\lambda_o + \lambda_d(a/\ell)^2$.[12,13] In the case of SB, we have both surface stress anisotropy and ledge-rebonding, thus we fit λ(SB) to $\lambda_o - \lambda_\sigma \ln(\ell/\pi a) + \lambda_d(a/\ell)^2$, with λ_σ taken to be the same as that in SA.[12,13]

From the present simulation results one can conclude that the elastic interactions on the stepped surfaces of Si(100) are of two origins, a relaxation due to stress anisotropy which manifests clearly in $\lambda(\ell)$ of structure SA, and a force-dipole interaction associated with ledge rebonding which accounts for the behavior of $\lambda(\ell)$ for structure DB. In the case of structure SB both effects are present. This conclusion is further supported by a direct analysis of the surface stress distributions in the three structural models.[12] By examining the stress profiles calculated using the relaxed and unrelaxed atomic configurations, one sees clearly the influence of stress relaxation and its corresponding effect on the strain energy.

We have also studied two vicinal surfaces, one composed of a sequence of alternating SA and SB ledges and another composed of DB. Qualitatively the behavior of a vicinal surface is similar to that of a flat surface.[12,13] Fitting the results using the same forms as above we find that λ_o(DB) is the same for the flat and vicinal surfaces, and that λ_o(vicinal) is equal to λ_o(SA)+λ_o(SB) for the flat surface. The coefficients of the stress-domain interaction λ_σ are also identical on the flat and vicinal surfaces. In contrast, the coefficients for dipole interaction λ_d are quite different, as might be expected since the moments induced by the ledges on flat and vicinal surfaces have opposite signs.

From the present simulations one finds that the single-layer stepped structure SA+SB is energetically favored at low miscut angle θ ($\tan\theta = S/L_y$, where S is the elevation of the vicinal surface over a distance L_y), whereas the double-layer structure DB is favored when θ exceeds about 1°.[12] Experimentally, in the temperature range 450 t0 550 K a crossover from double-layer to single-layer structures has been observed with θ in the region between 1° and 3.5°.[8,11] Since the simulation results above correspond to T = 0, one needs to consider the effects of thermal roughening of the ledges at finite temperatures.

A discrete-lattice model has been used to calculate the variation of the crossover angle θ_c with temperature.[11-13] As a result of thermal fluctuations SB ledges can form on SA stepped structure; however, even the lowest energy configuration, the rebonded SB ledge, requires an energy of ~ 60 meV/a. In contrast, the formation energy of a SA ledge on SB stepped structure is only about 6.3 meV/a. In the case of DB, formation of DA ledge due to roughening is unlikely because of the high energy cost. Thus, the roughening of SB is the only low-energy excitation that will be considered. The Hamiltonian we have analyzed is of the form [12,13]

$$H = \sum_i [\lambda_\perp |h_i - h_{i-1}| + \kappa h_i^2 + 2\epsilon_c(1 - \delta_{h_i,h_{i-1}})]$$

where h_i is the discrete-step fluctuation (in units of dimer length 2a) of the ith segment of the ledge front. In this model, the first term describes the strain energy associated with the stress domain interaction, with $\lambda_\perp$ being the energy cost per unit length of a ledge created in the perpendicular direction, and the second term describes the force-dipole interaction, with κ having contributions from λ_σ and λ_d. The

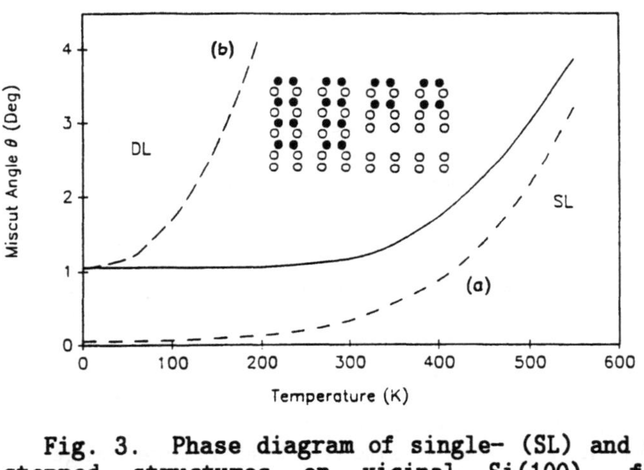

Fig. 3. Phase diagram of single- (SL) and double-layer (DL) stepped structures on vicinal Si(100), fluctuation model prediction based on simulation results for the various ledge energies (solid curve). Dashed curves denote the prediction of ref. 11 including a correction in the estimate of parameter κ (a), and the prediction without considering the corner-energy term in the Hamiltonian (b). Inset shows an atomic configuration with ledge corners, with solid and open circles representing atoms on the top and second layers, respectively.

third term describes the energy cost of creating a corner in a jagged ledge, with ϵ_c estimated to be 54 meV using the Stillinger-Weber potential. Assuming this model Hamiltonian the partition function has been calculated using the transfer-matrix method, and a phase diagram of the single- and double-layer stepped structures, shown in Fig. 3, derived.[12,13] One sees that the crossover miscut angle θ_c increases with temperature, indicating that the single-layer stepped structure SA+SB becomes more stabilized as thermal roughening sets in. The increase of θ_c by a factor of 3 at about 500 K leaves the theoretical prediction in good agreement with existing observations [8]. As another test of the present calculations, the magnitude of the fluctuating ledge, $[\langle h^2 \rangle]^{1/2}$, on a surface with $\theta = 0.5^\circ$ is calculated to be 50a and 60a at 400 K and 500 K respectively; this is also in agreement with observation [8,12,13].

In summary, we note that two types of elastic interactions have been found to play an important role in the equilibrium structures of Si(100) stepped structures. Stress anisotropy which occurs only on single-stepped surfaces gives rise to a relaxation varying logarithmically with the ledge separation ℓ, while ledge rebonding, present in both single- and double-stepped surfaces, shows a variation of ℓ^{-2}. While continuum elasticity predictions may appear to hold down to an essentially atomic level, we find that proper inputs from microscopic considerations are essential for a quantitative description of this problem.

4. Brittle-Ductile Transition in Fracture

An intrinsic property of a crystal is its fracture response under stress. The response is called brittle if an existing crack is able to propagate along a crystallographic plane in a cleavage manner; it is ductile if the crack-tip extension is accompanied by plastic deformation. While transitions from brittle to ductile behavior have been observed in a number of materials [15], this basic phenomenon, despite its technological interest, is not well understood in the mechanistic aspects of dislocation nucleation and mobility.[16]

We have carried out a molecular dynamics study of the behavior of a sharp crack in a uniaxially stressed crystal at finite temperatures.[17,18] A transition from brittle to ductile fracture occurring over a limited temperature range is directly observed. Detailed analyses of the atomic displacements reveal the transition mechanism to be the emission of dislocations from the crack tip. We find that the existing continuum description [16] is unable to account for the magnitude of the activation energy inferred from the transition temperature. We also find that thermal fluctuations and local stress effects have an important influence on the activation barrier.[17,19]

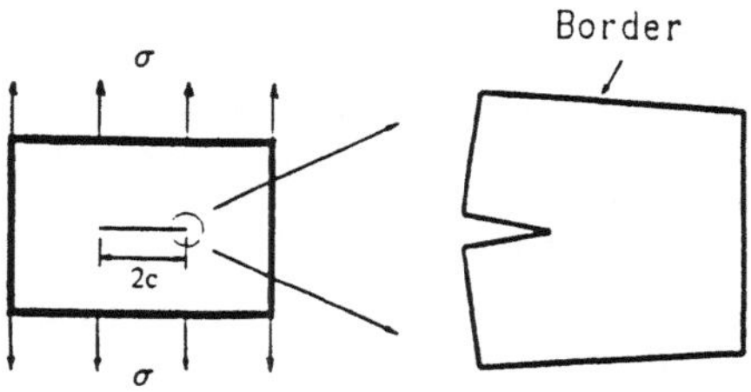

Fig. 4. Schematic drawing of a macroscopic body containing an atomically sharp crack of length 2c under an applied tensile stress σ (left hand figure) and the molecular-dynamics simulation cell representing only the crack-tip region with appropriate border conditions (right-hand figure).

The general setup of the simulation is illustrated in Fig. 4. We imagine a sharp crack embedded in a crystalline material of macroscopic dimensions which is subjected to uniaxial tension. The simulation cell of N atoms represents only the crack-tip region, as indicated, and the coupling between these atoms and the surrounding medium is expressed through a set of external forces imposed at the borders of the cell. These forces, determined at the outset and held constant during the simulation (in principle they could be updated periodically), represent the effects of surrounding medium in which atoms are displaced initially according to the linear-elasticity solutions to the crack-tip under stress. The advantage of this particular border condition is that the simulation cell, with the surrounding medium removed during the simulation, is free to deform in response to the stresses generated at the crack, thus allowing in a natural manner the nucleation and propagation of dislocations.

The potential function used in this study is an empirical potential constructed for the b.c.c. metal α-iron using an embedded-atom method.[21] It has been fitted to the experimental values of lattice parameter and sublimation energy precisely, and elastic constants and vacancy formation energy reasonably well. In our data analysis and interpretation we do not use experimental values for any of the physical parameters; for self-consistency all numerical values of physical properties are obtained directly from the model potential function.

We have examined several crack-tip configurations in the b.c.c. crystal lattice. Here we will discuss only the configuration specified as $[100](0\bar{1}1)$, the direction being that of the crack front and the plane is that in which the crack lies. The simulation cell is taken to be periodic in the direction along the crack front, hence the crack tip condition is that of plane strain.

A typical simulation run consists of setting up the crack tip configuration according to the elasticity solution for the atomic displacements [22] at a prescribed level of applied tensile stress, expressed as the quantity $K_I = \sigma\sqrt{\pi c}$, the stress intensity factor for mode I fracture. Here σ is the imposed stress and c the half-length of the macroscopic crack. The atoms are then allowed to move by molecular dynamics with the system temperature maintained at a constant value by rescaling the particle velocities after every time step of the simulation. If K_I is greater than the threshold value for the crack to propagate, interatomic bonds will begin to break at the crack tip [23]. On the other hand, if the stress is below the threshold level, one will see crack healing through the formation of new bonds.

In the case of brittle fracture a well-known criterion for the critical stress required for crack extension has been derived from an equilibrium condition between strain energy release rate and twice the surface energy of the crack surface.[24] For a crystal lattice in which anisotropic effects are taken into account, this so-called Griffith criterion K_{IG} has been obtained recently.[22] Using values of the compliance constants determined selfconsistently for the present potential function, we find that the Griffith prediction is accurate towithin about 5%, a significant improvement over previous studies which showed deviations ranging from 20% to a factor of 3. [20] We regard this as an indication of the physical meaningfulness of our simulation model, particularly with respect to treatments of the border condition and the anisotropy of the crack-tip structure.

To study the transition from brittle to ductile fracture we examine each crack tip at a supercritical stress of $1.1K_{IG}$ over a range of temperatures. In these runs the number of atoms in the cell ranges from 2400 to 3000, with corresponding cell dimensions ranging from 120A x 80A to 120A x 120A in the two directions shown in Fig. 5. At low temperatures brittle cleavage is observed in all cases. An example of cleavage fracture is given in Fig. 5(a) which shows the instantaneous atomic positions of the crack tip after 3000 time steps (4.4 psec) of simulation at T = 200 K. The atomic configuration of the same crack simulated at T = 400 K is shown in Fig. 5(b). Here one sees clearly the presence of two edge dislocations and a significant opening of the crack tip; both are characteristic features of ductile fracture.

The transition from brittle to ductile behavior as temperature is increased is also observed in the other two crack-tip orientations, [100](001), [110](001).[17] While the transition temperature is found to be somewhat different in each case, the fact that the transition occurs around room temperature is in agreement with available data.[25] To determine the deformation mechanism associated with the transition, a detailed analysis of the atomic trajectories is necessary. The simulation data from the [100](0$\bar{1}$1) crack model indicate that dislocations are emitted in the direction of the crack front [100]. From the trajectories of the atoms in adjacent (011) slip planes, one can discern appreciable movements in both the [100] and the [0$\bar{1}$1] directions, thus

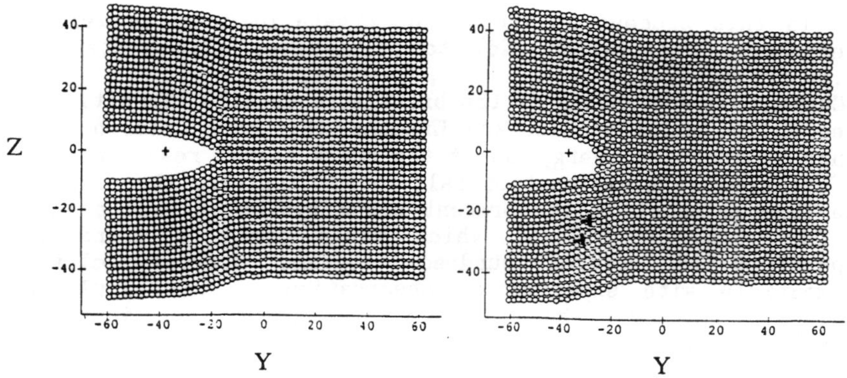

Fig. 5. Extension of [100](0$\bar{1}$1) crack, brittle fracture at T = 200 K (a, left-hand figure) and ductile fracture at T = 400 K (b, right-hand figure). Cross indicates initial crack-tip position.

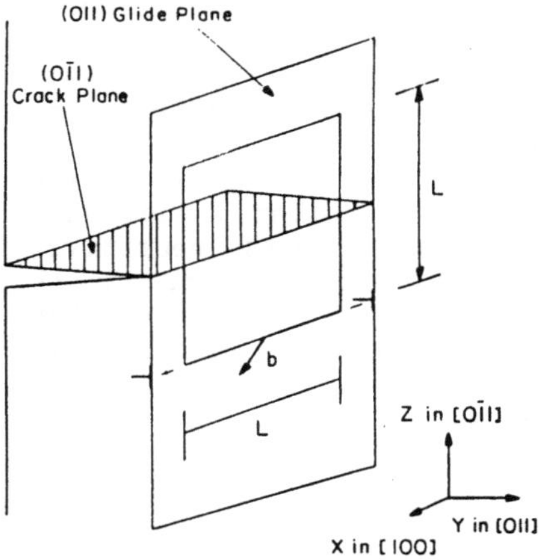

Fig. 6. Schematic drawing showing mechanism of ductile fracture by dislocation nucleation.

suggesting the presence of a screw component. We have developed a method for the quantitative characterization of dislocations through the use of an atomic disregistry function which is a measure of the relative displacement between corresponding atomic sites across the slip plane.[17] On this basis we find the dislocations nucleated from the crack. It is found that the dislocations emitted at the crack tip have mixed edge and screw components with Burgers vector b in the (1/2)[11$\bar{1}$] direction as shown in Fig. 6. The analysis therefore confirms that in the simulation dislocations are indeed nucleated from the crack tip and they move away by gliding on the (011) plane; the <111>{011} slip in this process is known from crystallography as well as experimental observation [26] to be a major slip system for b.c.c. crystals.

Current understanding of the brittle to ductile transition is based on the model of Rice and Thomson [16] which describes the change in the free energy ΔG of the crack system resulting from the creation of a dislocation half-loop. Applying the original Rice-Thomson model to the present crack system we find a value of $\Delta G \sim 80$ eV [17,19] which would imply a transition temperature of several hundred thousand degrees, clearly incompatible with experimental observation or the simulation findings. We have shown that it is possible to introduce several modifications of the Rice-Thomson formulation; these include taking into account the anistropy of the elastic constants and their softening as a result of temperature and local stress, as well as the nonlinear stress distribution at the crack tip.[17,19] These modifications give a significantly reduced value of ΔG, about 5 eV. Although such a barrier height is still too large to account for the observed transition temperature, this analysis clearly points to the important role of thermal and stress softening effects and that of nonlinear stress distribution. It is possible that other local processes may have to be taken into consideration in order to quantitatively account for the transition temperature revealed by the present simulations. In this respect it has been observed that during the simulation of crack-tip deformation and relaxation, local twinning can take place which suggests a kind of restricted shear transformation.[20,17] Other issues which warrant further studies include system size effects, sensitivity to details of the potential function, and updating of the traction forces at the simulation cell borders. Beyond these there lies the challenge of the problem of intergranular fracture, where one has the additional complexities associated with the presence of grain boundaries and solute or impurity segregation at the interfaces.[1]

5. Outlook

There is much evidence that the atomic-level understanding of solid surfaces and interfaces will become a major focus of materials research.[27] Because interfacial systems are inherently highly inhomogeneous (typical interfacial regions extend over a few atomic layers), their investigations require the use of techniques capable of atomic-scale resolution. While this poses considerable difficulties for experimental studies, it is actually an advantage for atomistic simulations and electronic-structure calculations.[1] In the present discussion we have demonstrated that simulation can provide results and details which are simply not obtainable by other means. In conclusion, it is important to keep in mind that ultimately it is the integrated approach where one combines simulation with macroscopic theory and experiment that holds the greatest promise for providing physical insight into the complex properties and behavior of materials systems.

Acknowledgments

The work discussed here has involved many collaborators, D. Wolf, S. Phillpot, and J. Lutsko in the melting studies, T. Poon, F. Abraham, and P. Ho in the Si(100) surface studies, and K. Cheung and A. Argon in the fracture studies. I am particularly indebted to D. Wolf for many stimulating and insightful discussions. Support by the Division of Materials Science, Argonne National Laboratory, through a Special Term Appointment, the IBM Thomas J. Watson Research Center through the IBM-MIT Joint Studies Program, and the U.S. National Science Foundation through Grants CHE-8806767 and CES-8815942 is gratefully acknowledged. Computational facilities were provided by the Argonne National Laboratory, the IBM Watson Research Center, and the San Diego Supercomputer Center.

References

[1] For an overview see MRS Bulletin, September and October 1990, introductions by D. Wolf and S. Yip, and article by S. R. Phillpot, D. Wolf, and S. Yip; see also S. Yip and D. Wolf, Mater. Sci. Forum 46, 77 (1989), and various articles in Atomic-Level Properties of Interface Materials, D. Wolf and S. Yip, eds. (Chapman and Hall, London, 1991), in press.
[2] A. R. Ubbelohde, Molten States of Matter: Melting and Crystal Structure (Wiley, Chichester, 1978).
[3] R. W. Cahn, Nature 323, 668 (1986); see also R. W. Cahn and W. L. Johnson, J. Mater. Res. 1, 724 (1986).
[4] S. R. Phillpot, J. F. Lutsko, D. Wolf, and S. Yip, Phys. Rev. B40, 2831 (1989), J. F. Lutsko, D. Wolf, S. R. Phillpot, and S. Yip, Phys. Rev. B40, 2841 (1989); see also S. R. Phillpot, S. Yip, and D. Wolf, Computers In Physics 3, No. 6, p. 20 (1989), and D. Wolf, P. R. Okamoto, S. Yip, J. F. Lutsko, and M. Kluge, J. Mater. Res. 5, 286 (1990).
[5] F. H. Stillinger and T. A. Weber, Phys. Rev. B31, 5262 (1985).
[6] J. Q. Broughton and X. P. Li, Phys. Rev. B35, 9120 (1987).
[7] D. J. Chadi, Phys. Rev. Lett. 59, 1691 (1987).
[8] See, example, A. J. Hoeven, J. M. Lessinck, D. Dijkkamp, E. J. van Loenen, and J. Dieleman, Phys. Rev. Lett. 63, 1830 (1989); M. Mundschau, E. Bauer, and W. Telieps, Philos. Mag. A61, 257 (1990); R. J. Hamers, U. K. Kohler, and J. E. Demuth, J. Vac. Sci. Technol. A8, 195 (1990).
[9] V. I. Marchenko, Sov. Phys. JETP Lett 33, 381 (1981).
[10] O. L. Alerhand, D. Vanderbilt, R. D. Meade, and J. D. Joannopoulos, Phys. Rev. Lett 61, 1973 (1988).
[11] O. L. Alerhand, A. N. Berker, J. D. Joannopoulos, D. Vanderbilt, R. J. Hamers, and J. E. Demuth, Phys. Rev. Lett. 64, 2406 (1990).

[12] T. W. Poon, Ph.D. Thesis, MIT (1990).
[13] T. W. Poon, S. Yip, P. S. Ho, and F. F. Abraham, Phys. Rev. Lett. 65, 2161 (1990).
[14] V. I. Marchenko and A. Ya. Parshin, Sov. Phys. JETP 52, 129 (1980).
[15] See, for example, G. T. Hahn, B. L. Averbach, W. S. Owen, and M. Cohen, in Fracture, B. L. Averbach et al., eds. (Technology Press, Cambridge, 1959), p. 91; P. B. Hirsch, S. G. Roberts, and J. Samuels, Proc. Roy. Soc. Lond. A421, 25 (1989).
[16] J. R. Rice and R. Thomson, Philos. Mag. 29, 73 (1974); see also A. S. Argon, Acta. Metall. 35, 185 (1987).
[17] K. S. Cheung, Ph.D. Thesis, MIT (1990).
[18] K. S. Cheung and S. Yip, Phys. Rev. Lett. 65, 2804 (1990).
[19] K. S. Cheung, A. S. Argon, and S. Yip, J. Appl. Phys., Feb. 15, 1991.
[20] B. deCelis, A. S. Argon, and S. Yip, J. Appl. Phys. 54, 4864 (1983).
[21] R. J. Harrison, A. F. Voter, and S. P. Chen, in Atomistic Simulation of Materials - Beyond Pair Potentials, V. Vitek and D. J. Srolovitz, eds. (Plenum Press, New York, 1989).
[22] G. C. Sih and H. Liebowitz, in Fracture: An Advanced Treatise, H. Liebowitz, ed. (Academic Press, New York, 1968), vol 2, p. 67.
[23] An interatomic bond across the crack plane is considered broken if the atom separation exceeds 1.5 times the distance in the equilibrium stress-free crystal; similarly, a bond is considered formed when the separation between a pair of atoms becomes less than this amount.
[24] A. A. Griffith, Phil. Trans. R. Soc. A221, 163 (1920).
[25] F. A. McClintock and A. S. Argon, eds. Mechanical Behavior of Materials (Addison-Wesley, Reading, 1966), p. 561.
[26] S. M. Ohr, Mat. Sci. Eng. 72, 1 (1985).
[27] Materials Science and Engineering for the 1990s: Maintaining Competitiveness in the Age of Materials, Report of the National Research Council, Committee on Materials Science and Engineering (National Academy Press, Washington, D.C., 1989).

Computer Experiments on Systems Containing Surfactants

T. Kawakatsu and K. Kawasaki

Department of Physics, Kyushu University 33, Fukuoka 812, Japan

Abstract. We review our recent results of computer experiments on the dynamics of systems containing surfactant molecules using a hybrid model, which is a combination of a continuum field and discrete molecules, proposed by the present authors. Phase separation processes of an immiscible binary mixture containing surfactant molecules and dynamics of surfactant bilayers, such as biomembranes or soap bubble films, are investigated and the results are compared with those of real experiments.

1. Introduction

It is well known that systems containing surfactant molecules show various complex domain structures on microscopic and mesoscopic scales [1,2]. The surfactant molecule is composed of two chemically distinct parts, which dissolve in different solvents. Such a feature is denoted as amphiphilic. Due to this amphiphilic nature, surfactant molecules are easily adsorbed onto interfaces between two immiscible solvents and reduce the surface tension, which permits interfaces to form complex structures on a mesoscopic scale of the order of 100 - 1000 Å. In oil-water-surfactant 3-component mixtures, we observe various domain structures, such as bicontinuous, micellar and lamellar structures, which will be discussed in this article. Biomembranes and soap bubble films are other examples of surfactant systems. They are formed by phospholipid molecules or soap molecules, which form double layers with opposite molecular orientations.

The molecular dynamics method would be a powerful tool to investigate dynamics of these surfactant systems microscopically. Up to now, however, the method has mainly been applied for investigating static or equilibrium aspects on microscopic scales of surfactant aggregates and surfactant bilayers [3-5], and there seem to be very few studies devoted to dynamical aspects on mesoscopic scales. Recently, we have proposed a hybrid model, which is suitable for investigating dynamics of surfactant systems on mesoscopic scales, and have performed computer experiments. In this article, we review recent results obtained in these investigations. For the details, readers may refer to the original papers [6-11].

2. The Hybrid Model

Let us consider surfactant molecules dissolved in a binary mixture composed of two components, A and B, which are mutually immiscible. For example, readers should consider an A-B binary polymer blend containing AB block copolymers, where the AB block copolymer is a polymer chain composed of an A-subchain and a B-subchain, which are connected at one end. Thus, the block copolymer has an amphiphilic nature.

In order to investigate characteristic features common to various surfactant systems, the model system should be as simple as possible but should retain the essential features of surfactant molecules, which are:

1) The surfactant molecule is composed of two chemically distinct parts, one A-philic and the other B-philic, and therefore has an amphiphilic nature.
2) Surfactant molecules are adsorbed onto interfaces of the binary mixture and reduce the surface tension.

We have proposed a hybrid model, which takes these features into account, where a continuous field and discrete molecules are coupled [6,7]. In our hybrid model, the binary mixture and the surfactant are treated in different manners. The binary mixture is described by a continuous scalar field $X(\boldsymbol{r},t)$, which is defined as the density difference between A and B components. On the other hand, surfactant molecules are treated as discrete molecules with director vectors, where the director vector is a unit vector along the direction from the B-philic center to the A-philic center of the surfactant molecule. Let $\boldsymbol{r}_i(t)$ and $\hat{\boldsymbol{s}}_i(t)$ be the position and the director of the i-th surfactant molecule at time t. The free energy functional of the system, H, can then be expressed as

$$H = H_{XX} + H_{XS} + H_{SS}, \tag{2.1}$$

$$H_{XX} = \int d\boldsymbol{r}[\frac{1}{2}D_X\{\nabla X(\boldsymbol{r})\}^2 - \frac{c}{2}X^2 + \frac{u}{4}X^4], \tag{2.2}$$

$$H_{XS} = \mu_S N_S + \frac{ql}{2}\sum_i \int d\boldsymbol{r} V_-(\boldsymbol{r} - \boldsymbol{r})\hat{\boldsymbol{s}}_i \cdot \nabla X(\boldsymbol{r}), \tag{2.3}$$

$$H_{SS} = q^2 \sum_{i<j}[2V_+(\boldsymbol{r}_{ij}) + \frac{l^2}{4}(\boldsymbol{s}_{ij}^- \boldsymbol{s}_{ij}^-) : \nabla\nabla\phi(\boldsymbol{r}_{ij}) + \frac{l^2}{4}(\boldsymbol{s}_{ij}^+ \boldsymbol{s}_{ij}^+) : \nabla\nabla\psi(\boldsymbol{r}_{ij})], \tag{2.4}$$

where D_X , c, u, μ_S, q and l are positive constants, N_S is the total number of surfactant molecules in the system, $\boldsymbol{s}_{ij}^{\pm} \equiv \hat{\boldsymbol{s}}_i \pm \hat{\boldsymbol{s}}_j$ and $\boldsymbol{r}_{ij} \equiv |\boldsymbol{r}_i - \boldsymbol{r}_j|$. Here, l is the measure of the linear dimension of the surfactant molecule, q the measure of the strength of the interaction of surfactant molecule, $\phi(\boldsymbol{r})$ and $\psi(\boldsymbol{r})$ are interaction potentials between surfactant molecules and the field X, where $\phi(\boldsymbol{r})$ is the attractive interaction between chemically similar parts

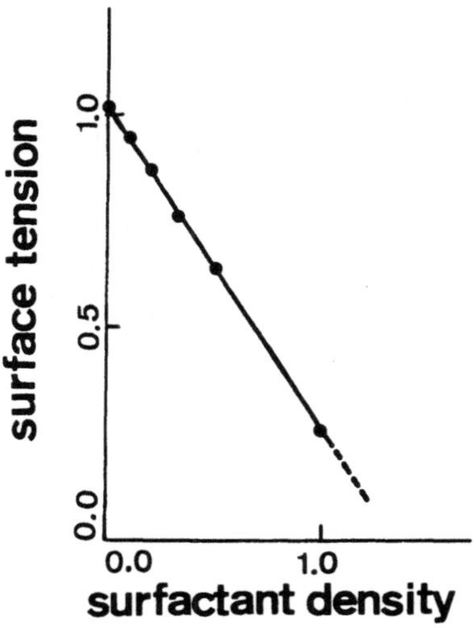

Fig.1 The surface tension of a surfactant-adsorbed interface as a function of the surfactant density on the interface.

and $\psi(\boldsymbol{r})$ is the repulsive interaction between chemically dissimilar parts, and $V_{\pm}(\boldsymbol{r}) \equiv \phi(\boldsymbol{r}) \pm \psi(\boldsymbol{r})$. We should note that, in deriving (2.1)-(2.4), we have neglected the excluded volume of the surfactant molecule and assumed that the A-B binary mixture is incompressible. In (2.1)-(2.4), H_{XX} accounts for the free energy functional of the phase separating binary mixture, which has the usual Ginzburg-Landau form, H_{XS} is the interaction between the field X and surfactant molecules and H_{SS} is the interaction between surfactant molecules, respectively. Terms of $O(l)$ and $O(l^2)$ represent director-field interactions and director-director interactions, respectively. The most important part is the coupling between surfactant molecules and the field X of the form $\hat{\boldsymbol{s}}_i \cdot \nabla X(\boldsymbol{r})$, which accounts for the amphiphilic nature of the surfactant molecule. In Fig.1, we show the surface tension of the surfactant-adsorbed interface as a function of the surfactant density on the interface, which is numerically obtained from (2.1)-(2.4) for 2-dimensional systems under the assumption that the surfactant density is low. It can be seen from Fig.1 that our hybrid model can successfully reproduce the reduction in the surface tension. Such a reduction in the surface tension originates mainly from the $\hat{\boldsymbol{s}}_i \cdot \nabla X$ term. Dynamics is introduced into this hybrid model by assuming purely dissipative equations of motion;

$$\frac{\partial}{\partial t} X(\boldsymbol{r}, t) = L^X \nabla^2 \frac{\delta H}{\delta X(\boldsymbol{r})}, \tag{2.5}$$

$$\frac{d}{dt} \boldsymbol{r}_i(t) = -L^{\rho} \frac{\partial H}{\partial \boldsymbol{r}_i}, \tag{2.6}$$

$$\frac{d}{dt} \hat{\boldsymbol{s}}_i(t) = -L^{s} [\frac{\partial H}{\partial \hat{\boldsymbol{s}}_i} - (\frac{\partial H}{\partial \hat{\boldsymbol{s}}_i} \cdot \hat{\boldsymbol{s}}_i) \hat{\boldsymbol{s}}_i], \tag{2.7}$$

where L^X, L^ρ and L^s are Onsager's kinetic coefficients and we have neglected cross couplings, thermal fluctuations and hydrodynamic effects. These evolution equations are integrated numerically with the use of the finite difference method for the field X (eq.(2.5)) and the molecular dynamics method for surfactant molecules (eqs.(2.6) and (2.7)), respectively. In the following, we restrict ourselves to 2-dimensional systems, for which we can take a much larger system size for computer experiments than for 3-dimensional systems. As the interaction potentials $\phi(r)$ and $\psi(r)$, we have employed exponentially decaying functions

$$\begin{aligned} \phi(r) &= -\alpha \exp(-r/\xi), \\ \psi(r) &= \beta \exp(-r/\xi), \end{aligned} \tag{2.8}$$

with $\alpha = 1.0$, $\beta = 5.0$ and $\xi = 1.0$, assuming that the interaction between the field and surfactant molecules is screened by the binary mixture.

3. Various Domain Structures

We first investigate the morphology of domain structures of A-B-surfactant 3-component mixtures [6,7]. Computer experiments have been performed as follows; The system is a square box with periodic boundary conditions on each side and contains the A-B binary mixture and N_S surfactant molecules. As an initial configuration for the computer experiments, we prepare a uniformly mixed state of the A component, the B component and surfactant molecules. Then we follow the temporal evolution of the system, which undergoes a phase separation process.

In Fig.2, we show typical domain structures observed in the late stage of phase separation processes [7]. Dotted regions and white regions represent A-rich domains and B-rich domains, respectively, and surfactant molecules are shown by small squares with tails, which represent both positions and directors of the individual surfactant molecules. We find (a) bicontinuous,

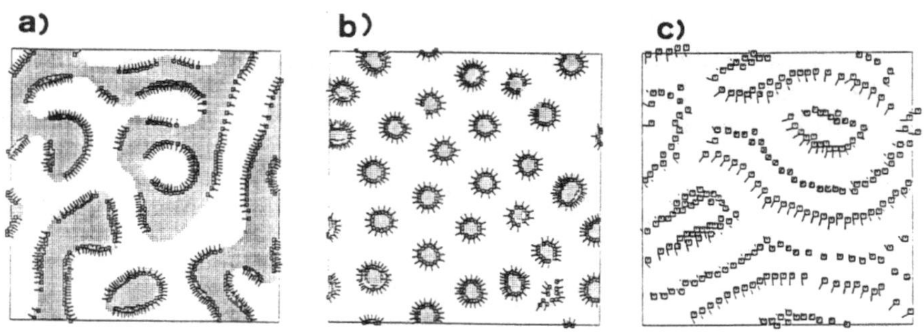

Fig.2 Typical domain structures observed in the late stage of phase separation processes. (a) bicontinuous, (b) micellar and (c) lamellar structures.

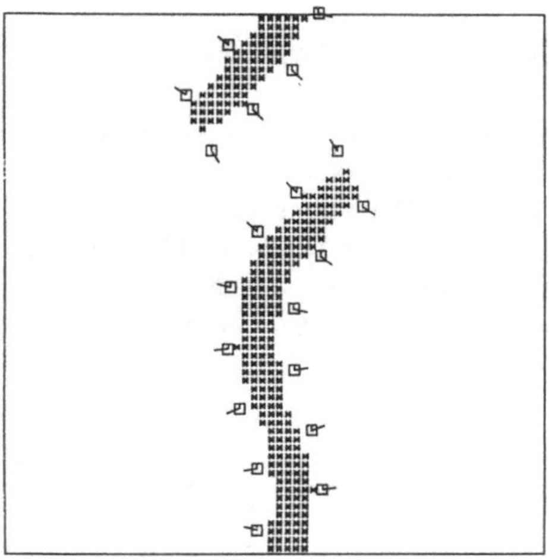

Fig.3 The surfactant bilayer formed by the hybrid model.

(b) micellar and (c) (imperfect) lamellar structures, all of which are actually observed in the real experiments on oil-water-surfactant mixtures. In these three cases, the fraction of the A component is (a) 50%, (b) 25% and (c) 0%, respectively, with the average number density of surfactant molecules, (a) and (b) $\bar{\rho}_S \equiv N_S/V = 0.347$ and (c) $\bar{\rho}_S = 0.694$, where V is the volume of the system.

Surfactant bilayers are special cases of lamellar morphology. In our hybrid model, a surfactant bilayer structure can also exist stably, which is shown in Fig.3 [12]. We will discuss such surfactant bilayers in section 5.

4. Phase Separation in a Binary Mixture Containing Surfactant Molecules

Now we investigate dynamical properties of phase separation processes in the presence of surfactant molecules in detail [7-9,11]. Here we discuss formation processes of the bicontinuous structure and the micellar structure shown in Figs.2(a) and (b).

4.1 Early Stage Dynamics

First we investigate dynamics in the early stage of the phase separation process of the bicontinuous structure [8-10]. In order to compare the results obtained by computer experiments with those of real experiments, it is convenient to calculate the scattering structure function of the field X, which is defined by

$$S_X(k,t) \equiv \frac{1}{2\pi}\int_0^{2\pi} d\theta \int d\boldsymbol{r} \exp(-i\boldsymbol{k}\cdot\boldsymbol{r}) < X(\boldsymbol{r},t)X(\boldsymbol{0},t) >, \quad (4.1)$$

where θ is the angle between the wave vector $\boldsymbol{k}$ and an arbitrarily chosen base direction and $< >$ denotes the average over different initial configurations. We show the temporal change of $S_X(k,t)$ for the bicontinuous case with the surfactant density $\bar{\rho}_S = 0.347$ in Fig.4(a), and also for $\bar{\rho}_S = 0.0$ in Fig.4(b) for comparison. When the surfactant is added (case (a)), the initial growth of S_X starts at higher wave number and with greater growth rate than in the

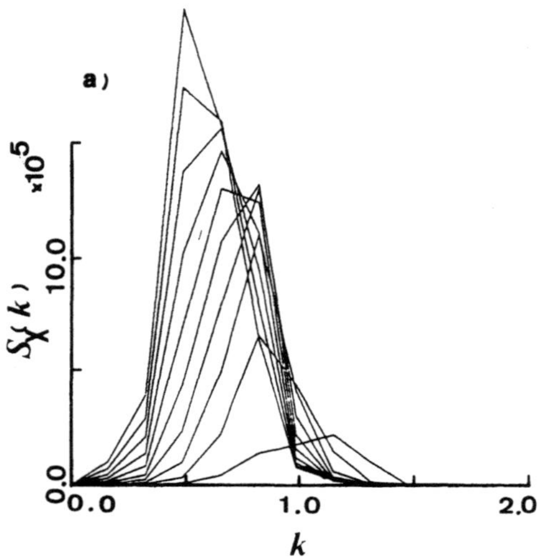

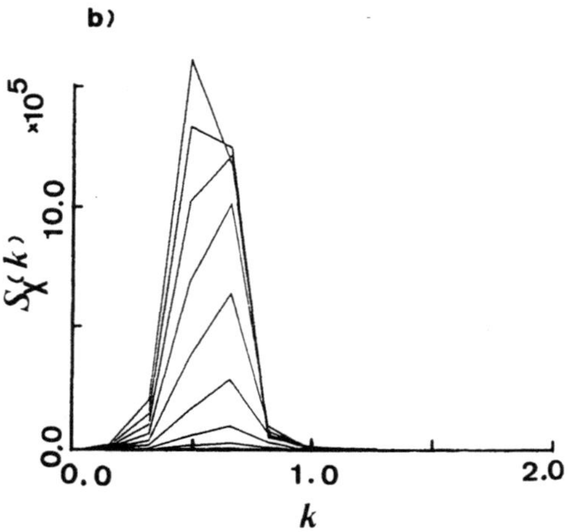

Fig.4 Temporal changes of $S_X(k,t)$ are shown for bicontinuous cases with (a) $\bar{\rho}_S = 0.347$ and (b) $\bar{\rho}_S = 0.0$, respectively.

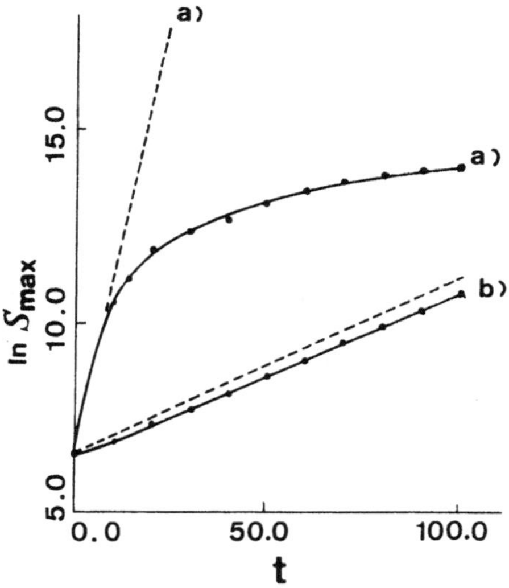

Fig.5 Temporal evolutions of $\ln S_X$ in the linear regime are shown for (a) $\bar{\rho}_S = 0.347$ and (b) $\bar{\rho}_S = 0.0$. Solid lines and broken lines show the results of computer experiments and linear analysis, respectively.

case without the surfactant (case (b)). For a more quantitative comparison, we show, by solid curves in Fig.5, the temporal changes of the logarithm of the peak height of $S_X(\boldsymbol{k},t)$, which is denoted as $\ln S_{\max}$, in the early stage for the two cases shown in Fig.4. In the absence of the surfactant, $\ln S_{\max}$ grows linearly in time, which corresponds to the well-known Cahn's linear regime, where $S_{\max} \sim \exp(\lambda t)$. On the other hand, when we add surfactant molecules to this system, the growth rate λ, which corresponds to the slope of the curve at $t = 0$, increases considerably. These results suggest that the phase separation is accelerated on a microscopic scale with the help of surfactant molecules.

In order to clarify the origin of this surfactant-assisted quick phase separation, we performed a linear analysis of fluctuations [9]. For the analytical treatment, it is easier to start from a full-continuous version of the hybrid model, where the surfactant is also represented by continuous field variables. Although a most general formulation has been given previously by the present authors [10], here we start from a rather simpler version, which can be derived directly from (2.1)-(2.7). Now, we define a director vector field of the surfactant by

$$\boldsymbol{s}(\boldsymbol{r},t) \equiv \sum_i \hat{\boldsymbol{s}}_i(t)\delta(\boldsymbol{r}-\boldsymbol{r}_i(t)), \tag{4.2}$$

which is a continuous field variable. Then, the hybrid equations of motion (2.1)-(2.7) can be rewritten as a set of continuum equations of motion. Retaining terms up to the first order in l and linearizing with respect to the small fluctuations around the initial uniform state, we obtain

$$\frac{\partial}{\partial t}\tilde{X}(\boldsymbol{k},t) = -L^X k^2[(D_X k^2 - c)\tilde{X}(\boldsymbol{k},t) + i\frac{1}{2}ql\boldsymbol{k}\cdot\{\tilde{V}_-(\boldsymbol{k})\tilde{\boldsymbol{s}}(\boldsymbol{k},t)\}], \tag{4.3}$$

$$\frac{\partial}{\partial t}\tilde{s}(\boldsymbol{k},t) = i\frac{1}{2}ql\bar{\rho}_S\ \frac{1}{d}[(d-1)L^s + L^\rho k^2]\boldsymbol{k}\tilde{V}_-(\boldsymbol{k})\tilde{X}(\boldsymbol{k},t), \tag{4.4}$$

where $\tilde{X}(\boldsymbol{k},t)$, $\tilde{s}(\boldsymbol{k},t)$ and $\tilde{V}_-(\boldsymbol{k})$ are the Fourier transforms of $X(\boldsymbol{r},t)$, $s(\boldsymbol{r},t)$ and $V_-(\boldsymbol{r})$, respectively. The first term in the square bracket on the r.h.s. of (4.3) is pertinent to phase separating simple binary mixtures without surfactant molecules. The second term on the r.h.s. of (4.3) and also the r.h.s. of (4.4) come from the coupling, $\hat{s}_i \cdot \nabla X$. Solving these coupled linear equations, we obtain temporal evolution of ln S_X in the linear regime and the results are shown in Fig.5 by broken lines. We find that the linear analysis can successfully reproduce the increase in the growth rate of $S_X(k,t)$ in the very early stage. We can also reproduce the peak shift of $S_X(k,t)$ to the higher wave number side quantitatively by this linear analysis, where the higher wave number originates from the interaction potential of the surfactant molecule, $V_-(\boldsymbol{k})$.

4.2 Late Stage Dynamics

We found that the growth of $S_X(k,t)$ is accelerated by the interaction $\hat{s}_i \cdot \nabla X$ in the early stage. Then, how about in the late stage ? In the late stage, the system is divided into A-rich domains and B-rich domains separated by surfactant-adsorbed interfaces, as is shown in Figs.2(a) and (b) [7,8]. The domain growth behavior can be seen by calculating the characteristic wave number $< k(t) >$ defined as the first moment of $S_X(k,t)$ as

$$< k(t) > \equiv \int_0^\infty dk\ kS_X(k,t) / \int_0^\infty dk S_X(k,t). \tag{4.5}$$

We show temporal changes of $< k(t) >$ for the bicontinuous case (Fig.2(a)) in Fig.6 and for the micellar case (Fig.2(b)) in Fig.7, respectively. In Fig.6, we also show the growth of the bicontinuous structure without surfactant molecules, where the usual $< k(t) > \sim t^{-1/3}$ behavior can be observed. When we add surfactant molecules, a slowing down of coarsening in the late stage can be seen for the micellar case in Fig.7. We can confirm from Fig.2(b) that interfaces are almost saturated by surfactant molecules at this stage. We cannot, however, observe the slowing down of coarsening for the bicontinuous case. This is because the interfaces are not yet saturated by surfactant molecules as can be seen from Fig.2(a), and the slowing down is also expected for this bicontinuous case at a much later stage. Clearly, the origin of such a slowing down in the late stage is repulsive interactions between surfactant molecules, which avoid the overlapping of two molecules and result in a reduction of the surface tension.

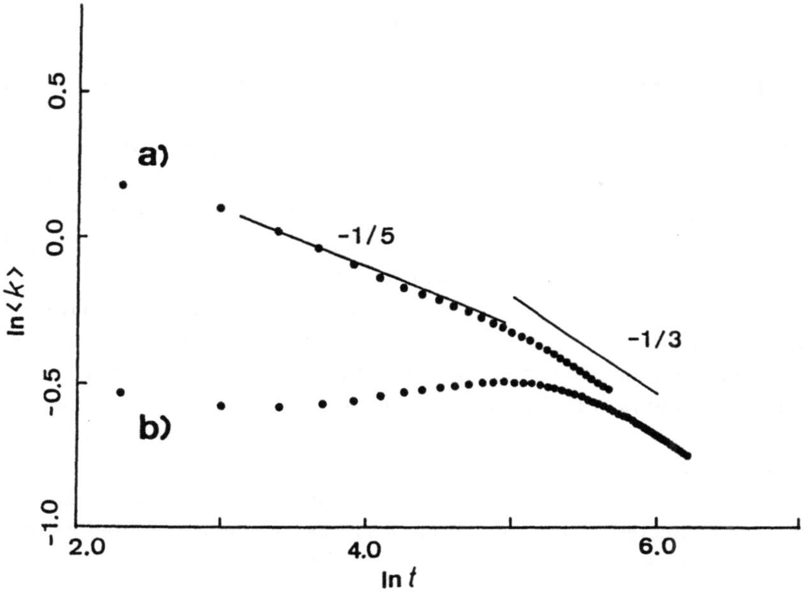

Fig.6 Temporal changes of the characteristic wave number $< k(t) >$ are shown for the bicontinuous case with (a) $\bar{\rho}_S = 0.347$ and (b) $\bar{\rho}_S = 0.0$, respectively.

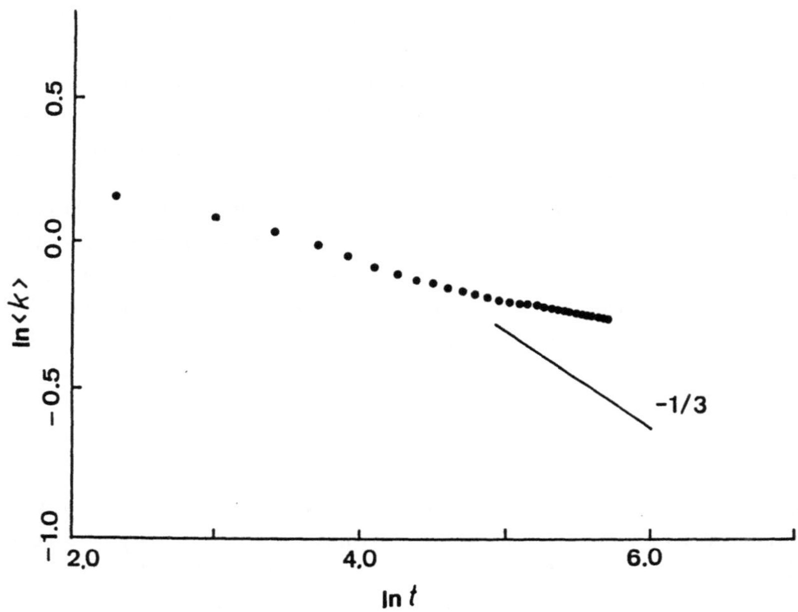

Fig.7 Temporal change of $< k(t) >$ for the micellar case is shown.

4.3 Comparison with Real Experiments

We expect that the results obtained in the preceding subsections are common to all immiscible binary mixtures containing surfactant molecules, because the coupling $\hat{s}_i \cdot \nabla X$ is expected to be universal over such systems. It is important to compare these results with the results of real experiments [9]. Up to now, very few experiments on real systems that directly correspond to our computer experiments mentioned above have been reported. One such experiment has been done by Izumitani and Hashimoto [13] on the phase separation process of an immiscible polymer blend containing an amphiphilic block copolymer, which plays the role of a surfactant. In the late stage, they observed slowing down of coarsening when the surfactant is added, which agrees with our observation. In the linear regime of the early stage, however, they also observed slowing down of the growth of the scattering function, which is in contradiction to our result.

We can attribute such a disagreement to the excluded volume effects, which we have neglected in deriving (2.1)-(2.4). Taking such excluded volume effects into account, we reformulated the linear theory to find that the growth rate of $\ln S_{\max}$ is expressed as

$$\lambda(\bar{\rho}_S) \simeq \lambda(\bar{\rho}_S = 0) + (\varepsilon + \varepsilon')\bar{\rho}_S, \tag{4.6}$$

for small $\bar{\rho}_S$, where the first term on the r.h.s. is the same as that in the case without surfactant molecules, ε comes from the $\hat{s}_i \cdot \nabla X$ type coupling and ε' accounts for the excluded volume effects. As was discussed in section 4.1, ε is positive, which accelerates the coarsening. On the other hand, we can show that ε' is negative both by analytical calculation under certain simplifing assumptions [9] and by another Monte Carlo simulation on an Ising ferromagnetic system containing amphiphilic molecules [11]. We also find that $\varepsilon \sim O(v_S^2)$ and $\varepsilon' \sim O(v_S)$, where v_S is the volume occupied by surfactant molecules in unit of molecular volume of the monomer of the binary mixture. Thus, it can be understood that the excluded volume effects should be important for systems containing small surfactant molecules, which is actually the case of the real experiment discussed above. We expect then that the acceleration of the coarsening in the early stage may be observed in experiments using large surfactant molecules.

5. Surfactant Bilayers

As the final example of our computer experiments using the hybrid model, we show surfactant bilayers [12]. First, we estimate the free energy of a surfactant bilayer with the use of the continuum version of the hybrid expression of the free energy functional, which was used in section 4.1. For simplicity, let us consider two parallel straight lines separated by the distance D. We

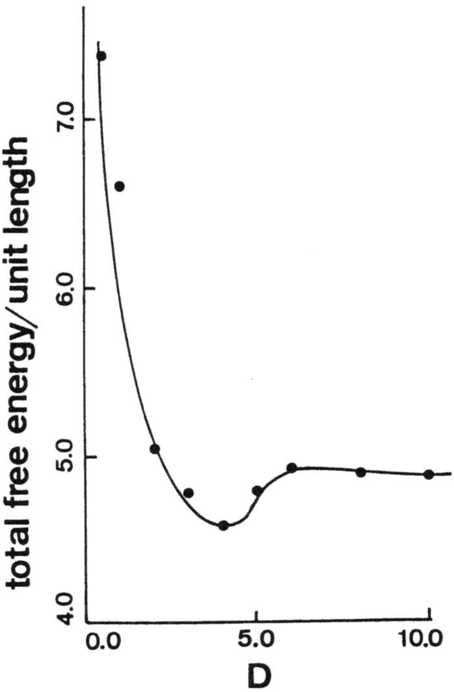

Fig.8 The total free energy of the surfactant bilayer is shown as a function of the separation distance between two layers, D at $\sigma = 0.5$.

assume that surfactant molecules are distributed uniformly on these two lines with opposite director vectors. We introduce the x-axis and the y-axis along the directions perpendicular and parallel to the layers, respectively. Then, from (2.3) and (2.4), we have

$$H_{XS} = 2L_y[\sigma\mu_S + \int_0^\infty dx X(x)\Omega(x)], \tag{5.1}$$

$$H_{SS} = -4q^2\sigma^2 L_y(\alpha-\beta)[DK_1(D)+1] + 2q^2l^2\sigma^2 Ly[\alpha V(D) - \beta V(0)], \tag{5.2}$$

where σ is the surface number density of surfactant molecules on each layer, L_y is the system length along the y-direction, α and β are defined in (2.8), $\Omega(x)$ and $V(x)$ are defined as

$$\Omega(x) \equiv -ql\sigma(\alpha+\beta)[(x+D/2)K_0(x+D/2)+(D/2-x)K_0(|x-D/2|)], \tag{5.3}$$

$$V(x) \equiv K_0(|x|) - |x|K_1(|x|), \tag{5.4}$$

and K_0 and K_1 are the modified Bessel functions of second kind. The equilibrium profile of the field X is obtained from the condition, $\delta H/\delta X(x) = 0$, which should be solved numerically under appropriate boundary conditions.

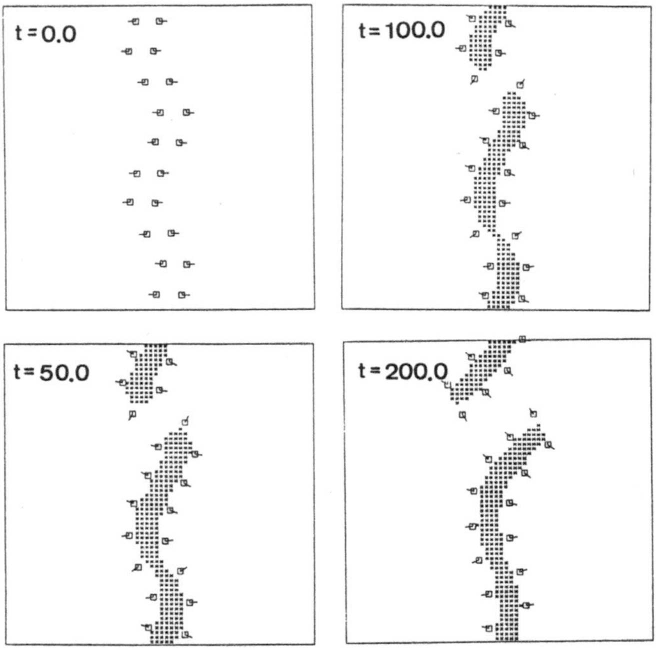

Fig.9 A relaxation process of an initially undulating bilayer is shown.

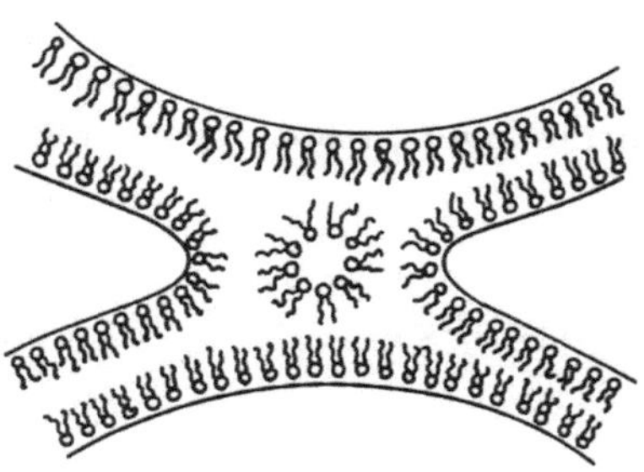

Fig.10 The transient structure formed in the course of a fusion of two bilayers is illustrated schematically.

In Fig.8, the total free energy per unit length as a function of D at a fixed value of σ is shown. We find a stable minimum around $D = 4.0$, which corresponds to the structure shown in Fig.3. We performed computer experiments on the relaxation processes of a bilayer, which is initially undulating, and one example is shown in Fig.9. In this case, the bilayer structure is maintained stably, although a fission of the bilayer takes place.

In order to investigate fusion processes of two bilayers, it should be necessary to incorporate thermal fluctuations into our computer experiments, because it needs some activation energy related with a transient structure, which is schematically illustrated in Fig.10. Such a computer experiment is a future problem.

6. Concluding Remarks

We have reviewed our computer experiments on surfactant systems. By these computer experiments, it has been shown that our hybrid model is applicable to various problems, such as phase separation processes and dynamics of surfactant bilayers. It should be emphasized that our hybrid model is simple enough to be treated also analytically, in contrast to the usual molecular dynamics computer experiment models on complex systems, where analytical treatment is often difficult due to the complexity of interaction potentials. As was mentioned in sections 4 and 5, in future studies, it would be important to incorporate excluded volume effects of surfactant molecules and thermal fluctuations in order to reproduce more faithfully the behavior of real systems. An extension to 3-dimensional systems and the inclusion of hydrodynamic effects are other challenging problems, where we could investigate diffusion processes of surfactant molecules on bilayers.

Acknowledgements

One of the authors (T.K.) would like to express his sincere gratitude to Dr. K.Sekimoto for fruitful discussions and suggestions. Discussions with Prof. T.Hashimoto, Mr. M.Takenaka, Dr. S.Tsuneyuki and the members of the laboratory are also acknowledged. Computations were carried out at the computer centers of National Institute for Fusion Science, National Laboratory for High Energy Physics, Institute for Molecular Science and Kyushu University. This work was partially supported by the Scientific Research Fund of the Ministry of Education, Science and Culture of Japan.

References

1) *Physics of Complex and Supermolecular Fluids*; eds. S.Safran and N.Clark (Wiley, New York, 1987).
2) *Modern Amphiphilic Physics*; eds. A.Ben-Shaul, W.Gelbert and D.Roux; to be published.
3) J.R.Gunn and K.A.Dawson; J. Chem. Phys. **91** (1989) 6393.
4) J.H.Hautman and M.L.Klein; J. Chem. Phys. **91** (1989) 4994.
5) K.Watanabe and M.L.Klein; J. Phys. Chem. **93** (1989) 6897.
6) T.Kawakatsu and K.Kawasaki; In *Strongly Coupled Plasma Physics, Proceedings of the Yamada Conference*, ed. S.Ichimaru (Elsevier Amsterdam, 1990).
7) T.Kawakatsu and K.Kawasaki; Physica **A167** (1990) 690.
8) T.Kawakatsu and K.Kawasaki; J. Colloid Interface Sci. **145** (1991) 413.
9) T.Kawakatsu and K.Kawasaki; J. Colloid Interface Sci. **145** (1991) 420.
10) K.Kawasaki and T.Kawakatsu; Physica **A164** (1990) 549.

11) T.Kawakatsu and K.Kawasaki; to be published in J. Colloid Interface Sci.
12) T.Kawakatsu and K.Kawasaki; unpublished work.
13) T.Hashimoto; private communication.

Collective Motion and Mechanism of Diffusion in Superionic Conductors

Y. Kaneko and *A. Ueda*

Department of Applied Mathematics and Physics, Faculty of Engineering, Kyoto University, Sakyo-ku, Kyoto 606, Japan

Abstract. The mechanism of collective diffusion in superionic conductor CaF_2 is investigated by the molecular dynamics method. The time of flight and the diffusion path are closely examined for individual F^- ions in a sequence of successive jumps. It is found that some of the F^- ions in a jump sequence have a long flight time and create temporal defects in the F^--sublattice. The sequences of correlated jumps are induced by these defects. The influence of the periodic boundary condition on the collective diffusion is also discussed.

1 Introduction

Transport properties of superionic conductors have important implications in condensed matter physics since they have both liquid-like properties of mobile ions and solid-like properties of the sublattice in the same phase.[1,2,3] In superionic materials the dynamic correlations among mobile ions as well as those between mobile and immobile ions play an important role for the diffusion. α-AgI is a prototype of the superionic conductors in which the behavior of mobile ions is typically liquid-like. Silver ions diffuse rather freely through the I^--bcc lattice since the number of the available sites (12(d) sites) is much larger than that of Ag^+ ions. In this system the correlation between mobile and immobile ions is important for diffusion. We have so far studied the dynamic correlations between Ag^+ and I^- through a series of molecular dynamics (MD) simulations [4,5,6], and proposed the mechanism of diffusion from a dynamical viewpoint.[7]

Fluorite compounds such as CaF_2 are typical of the superionic conductors which exhibit strongly correlated motions of mobile ions. Calcium ions form a closely-packed fcc lattice and fluorine ions occupy all the tetrahedral sites (*t*-sites). The octahedral sites (*o*-sites) in the fcc lattice provide the interstitial positions for diffusing F^- ions. However, the experimental [8] and simulational [9,10,11] studies showed that F^-

ions do not occupy octahedral voids in the superionic phase, but diffuse among the t-sites. In this case there is no empty site for mobile ions.

The purpose of this study is to elucidate the mechanism of collective diffusion in CaF_2 with use of a MD method. All the jumping motions of F^- between t-sites are picked out. We have followed the sequences of successive jumps and examined the time of flight and the diffusion path of individual F^- ions in a jump sequence. It is found that some of the F^- ions induce the jumps of neighboring F^- ions. Recently Doan *et al.* [12,13] discovered a new diffusion mechanism in CaF_2 and solid Na by the MD simulations. They illustrated that a ring sequence of atomic replacements is caused by an unstable Frenkel pair. We compare our results with those of Doan *et al.*

In the next section we outline our MD simulations. The results are given in Sec.3. After illustrating some patterns of correlated jumps, we present the results of the time of flight and discuss the mechanism of correlated jumps. The size dependence of the correlation is also discussed. Concluding remarks are given in Sec.4.

2 Molecular Dynamics Simulations

The MD simulations are performed for the 324-ion system (3×3×3 unit cells) and the 768-ion system (4×4×4 unit cells). The interionic potential is assumed as

$$\phi_{ij}(r) = \varepsilon(\frac{\sigma_i + \sigma_j}{r})^n + \frac{Z_i Z_j (fe)^2}{r}, \qquad (1)$$

where σ_i and Z_i are the ionic core radius and the valence of ion i, respectively, and f stands for the ionicity. The first term in Eq.(1) represents the soft-core repulsive potential and the second term the Coulomb potential. The potential parameters for CaF_2 are n=7, ε=0.28eV, σ_{Ca}=1.28A, σ_F=1.28A and f=1.0. The lattice constant is a=5.9A. As is shown in our previous papers [14,15], experimental results of the transition temperature and the diffusion constant of F^- are realized by this model. The equations of motion are integrated up to 38000Δt with the time mesh $\Delta t = 1.95\times10^{-15}$ sec. The average temperatures of the 324- and 768-ion systems are 1874K and 1752K, respectively, and the diffusion constants of F^- are 3.2×10^{-5} cm^2/sec and 3.7×10^{-5} cm^2/sec, respectively. The data from 5000Δt to 30000Δt are used in the following analysis.

3 Results

3.1 Correlated jumps of F^-

We have specified the positions of F^- ions by counting at every time step the number of F^- ions inside a sphere of an appropriate radius centered at each t-site. A movement of F^- between two spheres is regarded as a jump between t-sites. The radius of the sphere is taken to be 1/5 of the nearest neighbor distance between t-sites. In this way we obtained 4453 jumps for the 768-ion system and 1762 jumps for the 324-ion system. In both systems more than 80% of jumps occur in the $\langle 100 \rangle$ direction and about 15% in the $\langle 110 \rangle$ direction. These are in agreement with the earlier MD results of CaF_2 [9,10] and $SrCl_2$ [11].

In order to follow the sequence of successive jumps, we have numbered all the t-sites and F^- ions. When we get a jump of F^- from a site (site m) to the next site (site n), we search for a jump of another F^- from the site n. By repeating this procedure, we obtained the sequences of successive jumps. It is found that some of the sequences form a closed loop. The examples of the loops are shown in Fig.1. In Fig.1(a) the loop consists of three F^- ions, two ions moving in the $\langle 100 \rangle$ direction and one in the $\langle 110 \rangle$ direction. The loop in Fig.1(b) is made up of eight F^- ions. The jump sequence occurs on the (100) face of the three simple cubic cells. Most of the loops consist of 4-8 ions and occur within a few cells. Figure 2 illustrates the numbers of loops classified according to the number of their constituent F^- ions. Since most of the jumps occur in the $\langle 100 \rangle$ direction, many loops are composed of an even number of ions.

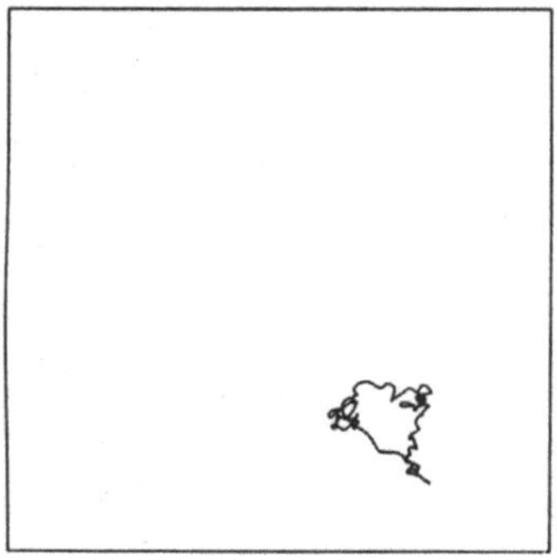

(a)

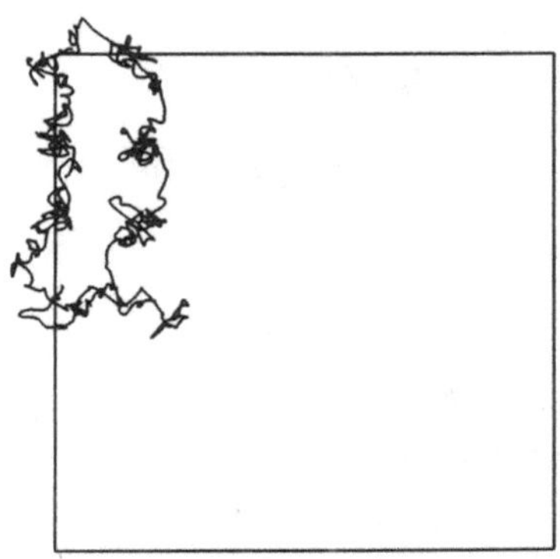

(b)

Fig.1 Examples of the loops of ionic jumps. The loop in (a) consists of three ions and in (b) eight ions.

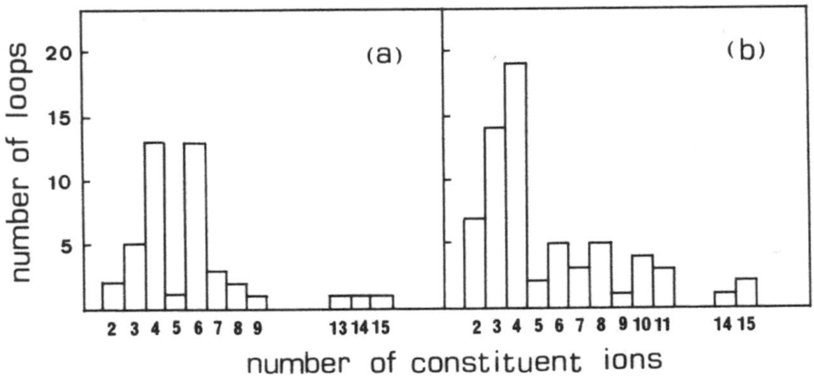

Fig.2 The numbers of loops and the numbers of their constituent ions for (a) the 324-ion system and (b) the 768-ion system. Figure (a) shows that there are two loops which consist of two ions, five loops which contain three ions, and so on.

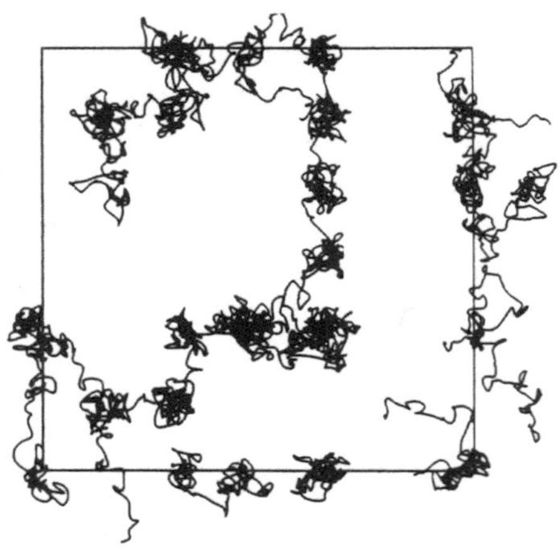

Fig.3 A long sequence of successive ionic jumps.

We also find long sequences of jumps. Figure 3 shows an example of a long jump sequence which does not form a closed loop. Each ion in the sequence moves from a t-site to the adjacent t-site and a jump of F^- always accompanies jumps of other F^- ions.

3.2 Time of flight

In order to investigate the characteristics of the correlated motions in more detail, we have evaluated the time of flight t_f of F^- ions. It is defined by the time required for F^- to move between t-sites. When F^- leaves a sphere at time t_1 and enters another sphere at time t_2, t_f is defined by $t_2 - t_1$. Figure 4(a) shows the flight times of four jumps which make up a loop. The lateral axis shows time and F^- ions are numbered as 132, 222, 152 and 224. The arrow is drawn from t_1 to t_2. The figures at the ends of the arrows stand for the numbers of the t-sites at which F^- stays before and after the movement.

This result provides us with crucial information about the diffusion mechanism. In Fig.4(a) the flight time of F^- 132 is longer than that of other F^- ions. After the movement of F^- 132, the jumps of F^- 224, 152 and 222 occur successively and their flight times are shorter than that of F^- 132. The trajectories of these ions are illustrated in Fig.4(b). The behavior of F^- 132 is different from those of other ions. The diffusion path of F^- 132 deviates largely from the shortest path between t-sites and deflects to the vicinity of an o-site, while the paths of other ions follow nearly a straight line connecting two sites. Therefore the particular motion of F^- 132 seems to induce the jumps of other F^- ions.

This diffusion mechanism is the same as that proposed by Doan *et al.* [12,13]. They studied by a MD method defects and disorders in CaF_2 assuming the Born-Mayer-Huggins potential. They demonstrated clearly that a ring sequence of atomic jumps is initiated by the formation of an unstable Frenkel pair, that is, a vacancy at t-site and an interstitial at

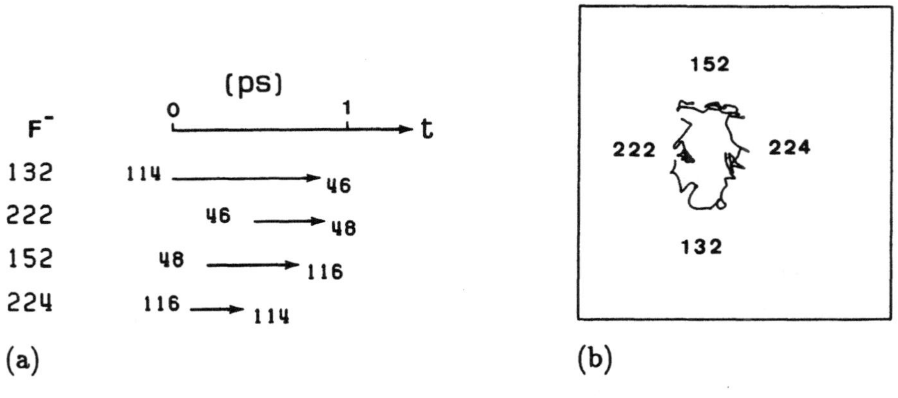

Fig.4 (a) The flight times of four ions. The numbers associated with t-sites are shown at the ends of the arrows. (b) The trajectories of F^- ions corresponding to (a).

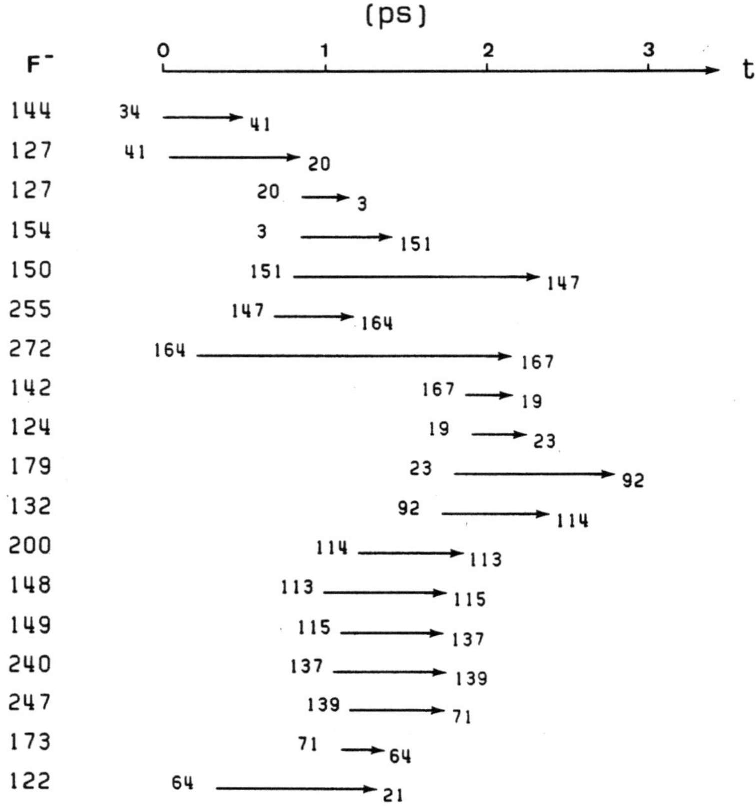

Fig.5 Flight times in a long sequence of jumps drawn in the same manner as in Fig.4(a).

o-site. Our result in Fig.4 is considered to show the same mechanism as that of Doan *et al.*

Figure 5 shows the flight times in a long sequence of jumps. The jumps with a long flight time and those with a short one appear repeatedly in the sequence. The jumps of F^- ions with a long flight time (F^- 150, 272 and 122) are always followed by the jumps of other F^- ions. This feature is also observed in other sequences. Therefore the driving mechanism shown in Fig.4 is considered to be an elementary process which gives rise to a long sequence of jumps. A small number of F^- ions move to the vicinity of *o*-sites and create temporal defects in the anion sublattice. The jumps of neighboring ions are induced by such defects. This process occurs from time to time and from place to place in the lattice, which results in a long sequence of correlated jumps.

3.3 Influence of the periodic boundary condition

In this section we shall discuss the size dependence of the correlated motions. Figure 6 shows the trajectories and the flight times of a jump sequence of six ions in the 324-ion system. As the 324-ion system consists of 3×3×3 unit cells, there are six t-sites aligned in the $\langle 100 \rangle$ direction. In Fig.6(a) six F^- ions situated at these sites move in the same direction and make up a ring sequence. One of the ions moves out from one side of the box and enters from the other side. This is clearly the effect of the periodic boundary condition. As shown in Fig.6(b), the jumps of six ions occur simultaneously and flight times are almost the same for all ions. The characteristic feature observed in Figs.4 and 5 is not observed in this case. Thus we should separate these events from our analysis since they arise from the effect of an artificial boundary condition. However, the small loops illustrated in Figs.1 and 4 do occur within the box and have no relation with the boundary condition. Therefore the diffusion mechanism discussed in the previous section is not affected by the size of the system.

4 Concluding Remarks

We have so far discussed the mechanism of correlated jumps in CaF_2. A ring sequence of ionic jumps is initiated by an unstable Frenkel pair as pointed out by Doan *et al.* Some of the F^- ions in a jump sequence move

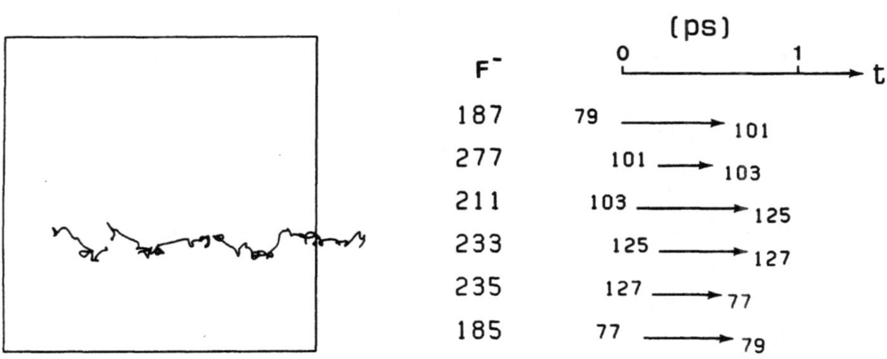

(a) (b)

Fig.6 (a) Jumps of six ions which make a ring sequence owing to the periodic boundary condition. (b) Flight times of F^- ions corresponding to (a).

to the vicinity of o-sites and create temporal defects in the anion sublattice. The jumps of surrounding ions are induced by such defects. This process occurs repeatedly in the system and gives rise to a long sequence of correlated jumps. This mechanism is considered to be irrelevant to the details of the interionic potential since our model with the potential (1) produces almost the same results as those of Doan *et al.*

Recently, it has been reported that the correlated jumps of atoms are observed in other systems such as glasses.[16] The result of the present study is expected to be a clue to elucidating the mechanism of collective motions in other fields.

REFERENCES

1. *Superionic Conductors*, ed. by G.D.Mahan and W.L.Roth, (Plenum, New York, 1976)
2. *Physics of Superionic Conductors*, ed. by M.B.Salamon, (Springer, Berlin, Heidelberg, 1979)
3. S.Chandra, *Superionic Solids* (North-Holland, Amsterdam, 1981)
4. A.Fukumoto, A.Ueda and Y.Hiwatari, J.Phys.Soc.Jpn.**51**, 3966 (1982)
5. M.Hokazono, A.Ueda and Y.Hiwatari, Solid State Ionics **13**, 151 (1984)
6. Y.Kaneko and A.Ueda, J.Phys.Soc.Jpn. **55**, 3924 (1986)
7. Y.Kaneko and A.Ueda, Phys.Rev.B **39**, 10281 (1989)
8. M.T.Hutchings, K.Clausen, M.H.Dickens, W.Hayes, J.K.Kjems, P.G.Schnabel and C.Smith, J.Phys.C **17**, 3903 (1984)
9. M.Dixon and M.J.Gillan, J.Phys.C **11**, L165 (1978)
10. G.Jacucci and A.Rahman, J.Chem.Phys. **69**, 4117 (1978)
11. M.J.Gillan and M.Dixon, J.Phys.C **13**, 1901,1919 (1980)
12. N.V.Doan and Y.Adda, Philos.Mag.A **56**, 269 (1987)
13. N.V.Doan, Philos.Mag.A **58**, 179 (1988)
14. Y.Kaneko and A.Ueda, J.Phys.Soc.Jpn. **57**, 3064 (1988)
15. Y.Kaneko and A.Ueda, *Proceedings of Yamada Conference XXIV on Strongly Coupled Plasma Physics*, ed. by S.Ichimaru, 631 (Elsevier, Amsterdam, 1990)
16. H.Miyagawa, Y.Hiwatari, B.Bernu and J.P.Hansen, J.Chem.Phys. **88**, 3879 (1988)

Index of Contributors

Springer Series in Solid-State Sciences

Editors: M. Cardona P. Fulde K. von Klitzing H.-J. Queisser

Springer Series in Solid-State Sciences

Editors: M. Cardona P. Fulde K. von Klitzing H.-J. Queisser

MIX
Papier aus verantwortungsvollen Quellen
Paper from responsible sources
FSC® C105338

If you have any concerns about our products,
you can contact us on
ProductSafety@springernature.com

In case Publisher is established outside the EU,
the EU authorized representative is:
Springer Nature Customer Service Center GmbH
Europaplatz 3, 69115 Heidelberg, Germany

Printed by Libri Plureos GmbH
in Hamburg, Germany